Nanoscale Physics for Materials Science

Takaaki Tsurumi • Hiroyuki Hirayama
Martin Vacha • Tomoyasu Taniyama

CRC Press
Taylor & Francis Group
Boca Raton London New York

CRC Press is an imprint of the
Taylor & Francis Group an **informa** business

CRC Press
Taylor & Francis Group
6000 Broken Sound Parkway NW, Suite 300
Boca Raton, FL 33487-2742

© 2010 by Taylor and Francis Group, LLC
CRC Press is an imprint of Taylor & Francis Group, an Informa business

No claim to original U.S. Government works

Printed in the United States of America on acid-free paper
10 9 8 7 6 5 4 3 2 1

International Standard Book Number: 978-1-4398-0059-1 (Hardback)

Library of Congress Cataloging-in-Publication Data

Nanoscale physics for materials science / Takaaki Tsurumi ... [et al.].
    p. cm.
    Includes bibliographical references and index.
    ISBN 978-1-4398-0059-1 (hardcover : alk. paper)
    1. Nanostructured materials. 2. Materials science. 3. Quantum theory. I. Tsurumi, Takaaki. II. Title.

TA418.9.N35N34555 2010
620.1'12--dc22   2009040942

Visit the Taylor & Francis Web site at
http://www.taylorandfrancis.com

and the CRC Press Web site at
http://www.crcpress.com

Contents

Preface vii
Acknowledgments ix
Authors xi

**Chapter 1   Fundamentals of quantum mechanics and
              band structure** 1
1.1   Fundamentals of quantum mechanics 1
    1.1.1   Probability amplitude and interference effects 1
    1.1.2   Uncertainty principle 5
    1.1.3   Wave functions 7
    1.1.4   Operators 8
    1.1.5   Eigenvalue and expected value 10
    1.1.6   Expansion theorem 10
    1.1.7   Schrödinger equation 11
    1.1.8   Principle of superposition 13
    1.1.9   Examples of solutions of the Schrödinger equation 14
        1.1.9.1   Electron in a one-dimensional (1D) box 14
        1.1.9.2   Harmonic oscillator 15
        1.1.9.3   Hydrogen atom 16
    1.1.10 Matrix mechanics and bra–ket (Dirac) notation 18
    1.1.11 Comparison of the Heisenberg and Schrödinger
        approaches to quantum mechanics 20
    1.1.12 Perturbation theory 22
1.2   Electronic band structure of solids 27
    1.2.1   Free electron Fermi gas 27
    1.2.2   Nearly free electron model (DOS) 31
    1.2.3   Bloch function 33
    1.2.4   Krönig–Penny model 33
    1.2.5   Tight binding model 35
    1.2.6   Phase velocity, group velocity, and
        effective mass 37

1.2.7   Reciprocal lattice and the Brillouin zone 40
1.2.8   Energy band structure of silicon (Si) 44
1.2.9   Tight binding approximation for calculating
        the band structure of graphene 45
1.2.10 Electron correlation 51
        1.2.10.1 Hartree–Fock approximation 51
        1.2.10.2 Density functional method 54
1.3   Material properties with respect to characteristic
      size in nanostructures 56
Problems 60
References 60

**Chapter 2   Electronic states and electrical properties of
              nanoscale materials** 63
2.1   Outline 63
2.2   Low dimensionality and energy spectrum 64
      2.2.1   Space for electrons in materials 64
      2.2.2   Electron DOS of 3D materials with macroscopic
              dimensions 65
      2.2.3   Electron DOS in 2D materials (nanosheets) 67
      2.2.4   Electron DOS in 1D materials (nanowires) 72
      2.2.5   Quantized conductance in 1D nanowire systems 74
      2.2.6   Electron DOS in 0D materials (nanodots) 77
2.3   Quantization 79
      2.3.1   2D square wells 80
      2.3.2   2D cylindrical wells 83
      2.3.3   Shape effect on the quantized states 85
      2.3.4   Finite potential wells 87
      2.3.5   Band dispersion effect 93
2.4   Edge (surface)-localized states 96
2.5   Charging effect 100
2.6   Tunneling phenomena 103
2.7   Limiting factors for size effects 111
      2.7.1   Thermal fluctuation 111
      2.7.2   Lifetime broadening effect 113
2.8   Electronically induced stable nanostructures 115
      2.8.1   Magic numbers in clusters 116
      2.8.2   Electronic growth 119
Problems 122
References 123

**Chapter 3   Optical properties and interactions of
              nanoscale materials** 125
3.1   Size-dependent optical properties: Absorption
      and emission 125

3.1.1   Basic quantum mechanics of linear
optical transitions 126

3.1.2   General concept of excitons............................. 133

3.1.3   Wannier excitons.. 135

3.1.4   Size effects in high-dielectric-constant materials 136

3.1.5   Size effects in π-conjugated systems.................... 140

3.1.6   Strongly interacting π-conjugated systems:
A molecular dimer... 144

3.1.7   Molecular Frenkel exciton 149

3.1.8   Size effects in molecular excitons: Coherence
length and cooperative phenomena..................... 153

3.1.9   Effects of finite number of optical electrons 157

3.2   Size-dependent optical properties: Absorption
and scattering.. 158

3.2.1   Basic theory of light scattering 160

3.2.2   Size-dependent scattering from dielectric
spheres: Mie solutions..................................... 164

3.2.3   Optical properties of metal nanoparticles:
Plasmonics 169

3.2.4   Local field enhancement and surface-enhanced
Raman scattering..176

3.3   Size-dependent electromagnetic interactions:
Particle–particle 179

3.3.1   Radiative energy transfer 179

3.3.2   Förster resonant energy transfer (FRET)............... 180

3.3.3   Electron-exchange (Dexter) energy transfer........ 187

3.3.4   Photo-induced electron transfer......................... 190

3.4   Size-dependent interactions: Particle–light interactions
in finite geometries... 191

3.4.1   Optical interactions in microcavities 191

3.4.2   Effects of dielectric interfaces 198

Problems.. 201

References.. 204

**Chapter 4   Magnetic and magnetotransport properties of
nanoscale materials** 207

4.1   Fundamentals of magnetism 207

4.1.1   Magnetic ions and magnetic ordering................ 207

4.1.2   Exchange interaction 208

4.1.3   Mean field theory of ferromagnetism.................211

4.2   Size and surface effects in 3D confined systems............ 213

4.2.1   Quantization of electronic structures and
the Kubo effect 214

4.2.2   Surface magnetism of transition noble metals 220

4.2.3   Single-domain structures and superparamagnetism 224

4.3   Ferromagnetic domain-wall-related phenomena 229
      4.3.1 Macroscopic quantum tunneling in magnetic
             nanostructures 229
      4.3.2 Electron scattering at domain walls: Quantum
             coherence.. 233
      4.3.3 Spin current and spin transfer torque–current-
             induced domain wall motion.. 235
4.4   Spin transport in magnetic nanostructures: Magnetic
      interface effect 240
      4.4.1 GMR and TMR effect: Spin-dependent scattering
             in multilayers and tunneling junctions.............................. 240
      4.4.2 Spin accumulation and current-perpendicular-to-plane
             (CPP) GMR: Spin diffusion length 245
      4.4.3 Spin Hall effect: Side jump and skew scattering
             due to spin–orbit coupling... 249
Problems.. 253
References.. 253

Index 257

Preface

Nanotechnology has been one of the driving forces of science and technology over the past 20 years. There is an abundance of monographs, textbooks, and related literature covering nanoscale science and technology from a wide range of perspectives. However, most publications focus on the *technology* of this field. There is no shortage of books describing the preparation, properties, and characterization of nanomaterials, with many books being devoted to specific nanomaterials, such as carbon nanotubes. On the other hand, there are few examples of texts giving a general cross-disciplinary description of the physical *phenomena* that govern the novel properties of nanomaterials. This was the motivation for writing this textbook, which is designed to complement a new graduate school course for materials science engineers at the Tokyo Institute of Technology. The course is an integral part of the Global Center of Excellence (G-COE) project and covers fundamental cross-disciplinary concepts in materials science and engineering. *Size dependence* is the keyword of this book, in which we describe physical phenomena that undergo qualitative or quantitative changes as the size of physical objects decreases. Many of these phenomena are not new; some were discovered in the early part of the twentieth century. Our goal in this book is to look at these phenomena from a new perspective by linking them with recent scientific and technological developments. The most dramatic physical changes occur on scales where the quantum nature of objects starts dominating their properties—in the range 0.1–1.0 nm—even though long-range electromagnetic interactions from 10 to 100 nm can be important in determining many properties. These scales define the title of the book: *Nanoscale Physics*.

The aim of the book is to give scientists and engineers of materials science a comprehensive description of the phenomena and changes that can be expected when macroscopically sized materials are reduced down to the nanometer level. Where possible, we have avoided the traditional division of materials science into inorganic, organic, semiconductor, ceramics, metallurgic, and so on, and instead kept our approach general. Thus, the book is divided according to physical phenomena and interactions. The first chapter is a review of the theoretical background that is necessary for understanding the remaining chapters. The next three chapters address the electrical, optical, and magnetic properties as functions of size and distance. For a deeper understanding of the contents of chapters, we have included a set of problems at the end of each chapter. This book is primarily intended for graduate school students affiliated with materials science and engineering departments. Such students will already have a good background in general physics and chemistry, will be conversant with the basics of quantum mechanics, and many will have some knowledge of solid-state physics, physical chemistry, or optics. The textbook should be an excellent source of reference material on phenomena not covered by standard courses on materials science.

Takaaki Tsurumi
Hiroyuki Hirayama
Martin Vacha
Tomoyasu Taniyama

Acknowledgments

Writing this book meant that we spent less time than usual with other people on other activities, both professionally and privately. We would like to thank all concerned for their understanding.

The book was written as part of the Global Center of Excellence (G-COE) project *Education and Research Center for Materials Innovation* at the Tokyo Institute of Technology, and the support from the G-COE is gratefully acknowledged.

The manuscript was carefully proofread and rewritten where necessary into naturally sounding English by Professor Adarsh Sandhu of the Quantum Nanoelectronics Research Center at the Tokyo Institute of Technology. Mieko Ozaki helped us with the technical aspects of the manuscript, and students from Professor Tsurumi's group redrew the illustrations with care. We thank all for their contributions.

Authors

Takaaki Tsurumi is a professor in the Department of Metallurgy and Ceramics Science at the Tokyo Institute of Technology, Tokyo, Japan. He obtained his doctorate degree from the Tokyo Institute of Technology in 1985 for his work on the crystal chemistry of oxide protonic conductors. He continued his research on oxide conductors while working as a postdoctoral fellow at McMaster University, Canada, from 1985 to 1986. After joining the Tokyo Institute of Technology as a research associate, he changed his field of research to dielectric and ferroelectric materials. He is currently studying the dielectric properties of oxide artificial superlattices, the structure–property relationship of piezoelectric ceramics, wide-range dielectric spectroscopy, and the electro-optical effect of perovskite-type thin films.

Hiroyuki Hirayama is a professor in the Department of Materials Science and Engineering at the Tokyo Institute of Technology, Tokyo, Japan. He studied at the University of Tokyo, Japan, receiving his doctorate degree in 1986 for research on the angle-resolved electron energy loss spectroscopy of In/Si(111) surfaces. Afterward, he joined NEC Corporation, where he succeeded for the first time in the development of gas source Si and Ge molecular beam epitaxy and its application to the fabrication of high-speed Si and Si/Ge transistors. From 1990 to 1991, he was a visiting scientist at the Fritz-Harber-Institut der Max-Planck-Gesellschaft, Berlin, with Professor G. Ertl, and studied laser-induced surface dynamics. He moved to the Tokyo Institute of Technology in 1995 to continue his work on surface and interface physics. His main research interest includes the control of the size and electronic states of nanostructures at solid surfaces and interfaces.

Martin Vacha is an associate professor in the Department of Organic and Polymeric Materials at the Tokyo Institute of Technology, Tokyo, Japan. He studied at Charles University in Prague, Czech Republic, where he also obtained his PhD degree in 1991 for work on the low-temperature, high-resolution optical spectroscopy of photosynthetic systems. He has extensive research experience in the fields of hole-burning and single-molecule spectroscopy of organic molecules and molecular complexes gained during stays at academic and government research institutions in Japan. He has held his current position at the Tokyo Institute of Technology since 2004. His main research interests include the nanoscale physical properties of organic materials and biomaterials studied by single-molecule techniques.

Tomoyasu Taniyama is an associate professor in the Materials and Structures Laboratory at the Tokyo Institute of Technology, Tokyo, Japan. He obtained his PhD degree from Keio University, Japan, in 1997 for his work on the magnetic properties of transition metal nanoparticles. He joined the National Research Institute for Metals to pursue the magnetotransport properties of lithographically defined magnetic nanostructures. In 1998, he moved to the Tokyo Institute of Technology. During his two-year stay at Cavendish Laboratory, University of Cambridge, from 2001 to 2003, he studied the transmission of optically excited spin-polarized electrons across ferromagnetic metal/semiconductor interfaces. His research is focused on the physical properties of spin electronic devices, in particular, spin injection, spin detection, and spin manipulation in magnetic heterostructures.

chapter one

Fundamentals of quantum mechanics and band structure

Chapter 1 reviews the theoretical principles necessary for understanding the remaining chapters of the book and also reviews the fundamentals of quantum mechanics and electronic band structures. Physical properties of materials on nanometer scale are governed by quantum mechanics. The first part of this chapter describes the fundamentals of quantum physics, including the uncertainty principle, the Schrödinger equation, matrix mechanics, and perturbation theory. The electronic structure of materials is determined by crystallographic periodicity, which is represented by the electronic band structure. In the second part of this chapter, the concept of electronic band structure is examined based on the nearly free electron approximation and the tight binding model. As an introduction to other chapters in the book, the final part of this chapter describes the effect of size on the electronic, optical, and magnetic properties of materials, with an emphasis on the nanometer scale and properties of nanomaterials.

1.1   Fundamentals of quantum mechanics

1.1.1   Probability amplitude and interference effects

We start with the famous double-slit experiments—the so-called *thought experiments*—employed by Richard Feynman to illustrate the difficult concepts of probability amplitude and interference in quantum mechanics [1].

1. *Double-slit experiment 1:* Probability of finding bullets reaching a wall. Imagine that we are facing a wall with a machine gun and there are two slits—1 and 2—between us and the wall (Figure 1.1). We shoot the machine gun toward the slits and some of the bullets reach the wall after colliding with the slits. Importantly, the interaction with the slits will change the velocity and direction of the bullets. Now, we close slit 2 and fire bullets toward slit 1, and after firing, we go to the wall to inspect the distribution of the bullets that have reached

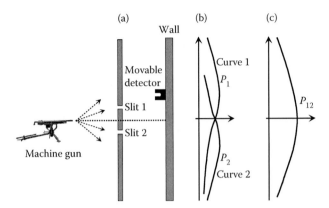

Figure 1.1 Double-slit experiment using bullets. (a) A view of the experimental setup, (b) probabilities of finding bullets on the wall obtained with either slit 2 (curve 1) or slit 1 (curve 2) closed, and (c) the probability of finding bullets on the wall obtained with both slits open.

the wall. We will find the distribution of bullets as shown in curve 1 in Figure 1.1b. Here, curve 1 shows the probability of finding bullets on the wall. The same experiment conducted with slit 1 closed will yield curve 2. If we carried out the same experiment with both slits open, then the probability of finding bullets on the wall will be given by the addition of curves 1 and 2, as shown in Figure 1.1c. It is easy from personal experience to envisage these results in the real world. The probability of finding bullets on the wall is given by

$$P_{12} = P_1 + P_2, \tag{1.1}$$

where P_{12} is the probability when both slits are open and P_1 and P_2 are the probabilities when either slit 2 or slit 1 is closed, respectively.

2. *Double-slit experiment 2:* Intensity of waves of water reaching a wall. In the next experiment, the machine gun is replaced by a wave generator for producing waves of water. We can make a simple wave generator by throwing small stones one by one, at a constant period to the same position onto the surface of water. Here, slits 1 and 2 are placed between the wave generator and the wall, as shown in Figure 1.2a. The intensity of the resultant wave, which is given by the square of the amplitude of the wave, is measured on the wall. The intensity of the resultant wave reaching the wall is proportional to the energy of the wave. At first, we make waves with slit 2 closed and measure the intensity of the wave reaching the wall. The variation of the wave intensity with the position of the wall is shown by curve 1 in Figure 1.2b, which is the same as in Figure 1.1b. Experiments with slit 1 closed will yield curve 2. However, experiments with both slits open will

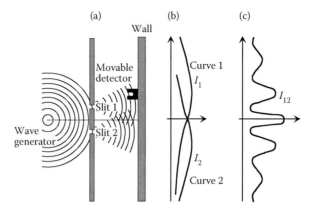

Figure 1.2 Double-slit experiments using waves of water. (a) A view of the experimental setup, (b) wave intensities on the wall obtained with either slit 2 (curve 1) or slit 1 (curve 2) closed, and (c) wave intensity on the wall obtained with both slits open.

give the intensity distribution shown in Figure 1.2c, which is totally different from the results obtained for experiments using bullets (Figure 1.1c). The change of the wave intensity with position on the wall is due to the so-called *interference* between the two waves passing through slits 1 and 2. The intensity I_{12} on the wall is given by

$$I_{12} = |A_{12}|^2 = |A_1 + A_2|^2 \neq |A_1|^2 + |A_2|^2 = I_1 + I_2, \tag{1.2}$$

where $|A_{12}|$ is the resulting amplitude when both slits are open and $|A_1|$ and $|A_2|$ are the amplitudes when either slit 2 or slit 1 is closed, respectively. Again, these results are easy to imagine in the real world.

3. *Double-slit experiment 3:* Probability of finding electrons reaching a wall.

Now let us move on to a world governed by quantum mechanics by replacing the wave generator with an electron gun, and the experiment is carried out in a vacuum. The electron gun consists of a tungsten wire and a metal electrode. The tungsten wire is heated to generate thermal electrons and the metal electrode is used to accelerate electrons by applying a positive voltage. Again, two slits, 1 and 2, are placed between the electron gun and the wall as shown in Figure 1.3a. A Geiger counter is placed on the wall to measure the number of electrons reaching the wall. The position of the counter can be changed up and down along the wall to determine the distribution of electrons reaching the wall. In the first experiment, slit 2 is closed—the same as in the experiments with the machine gun and waves of water. Electrons are not visible but we will hear a sound

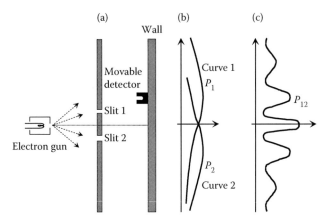

Figure 1.3 Double-slit experiments using electrons. (a) A view of the experimental setup, (b) probabilities of finding the electrons on the wall obtained with either slit 2 (curve 1) or slit 1 (curve 2) closed, and (c) the probability of finding the electrons on the wall obtained with both slits open.

"click" from the counter when an electron reaches the wall and enters the counter. It is very important that the sounds from the counter always have the same intensity, and we never hear weak or strong sounds. This constant sound implies that electrons are arriving one by one, like bullets. The interval between the sounds depends on the position of the counter on the wall. In this experiment, we count the sound over a certain period and plot the variation of the frequency of the sound with the position of the counter on the wall. We will get the results shown by curve 1 in Figure 1.3b, which is the same as curve 1 in Figures 1.1b and 1.2b. Experiments with slit 1 closed will give curve 2. We will interpret these results as implying that electrons behave like a particle—just like bullets. However, we will change our mind when we open both the slits. These experiments will show that the frequency of the sound is given by Figure 1.3c, which is consistent with the results for the waves on water, namely that electrons show interference effects. These experiments mean that electrons behave like particles as well as waves, a conclusion that is not acceptable in the real world. Before considering this phenomenon, let us represent these experimental results mathematically. The curves in Figure 1.3b and c can be regarded as the probability of finding electrons at different positions on the wall. We should regard the probability as the intensity of waves as in the case of waves of water. These assumptions lead to the following expression:

$$I_{12} = |A_{12}|^2 = |A_1 + A_2|^2 \neq |A_1|^2 + |A_2|^2 = I_1 + I_2, \qquad (1.3)$$

where I_{12} is the probability of finding electrons, $|A_{12}|$ is the *probability amplitude* obtained with both slits open, and $|A_1|$ and $|A_2|$ are the probability amplitudes when either slit 2 or slit 1 is closed, respectively.

1.1.2   Uncertainty principle

In the double-slit experiment 3, conducted using an electron gun, we will encounter an intriguing question: Which of the two slits do the electrons pass through when both slits are open? To solve this mystery, let us continue the double-slit experiments a little further by putting a strong light source near the two slits as an electron detector as shown in Figure 1.4a. The charge of an electron scatters the light that is observable with the naked eyes as a flash of light. In this experiment, we can see flashes near slit 1 or slit 2 when electrons pass through either of the slits, and we can also detect electrons with a Geiger counter. If the experiment is successful, we will know which slit electrons pass through when both slits are open. However, we will discover remarkable results that show that the probability of finding electrons changes according to the curve shown in Figure 1.4b. These results show that detection of the positions of the electrons causes the removal of the interference-related effects of the electrons from the experimental results. Namely, under these circumstances, the electrons behave like bullets.

We may have realized that there is a fault in this experiment. The energy of an electron changed when it collided a photon with an energy of $h\nu$, where h is Planck's constant and ν is the frequency of the light. This change in energy may affect the probability distribution of finding

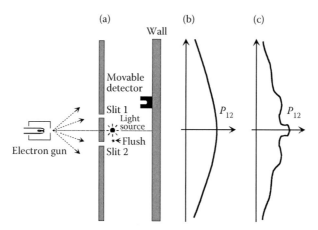

Figure 1.4 Double-slit experiments using a light source to detect electrons. (a) A view of the experimental setup, (b) probabilities of finding the electrons on the wall obtained with a light source with a short wavelength, and (c) probabilities of finding the electrons on the wall obtained with a light source with a long wavelength.

electrons on the wall. To overcome this, in the next experiment, we could increase the wavelength of the light—the energy of photons $h\nu$ is reduced by increasing their wavelengths (decreasing frequency)—so as not to disturb the probability of finding electrons on the wall. In such an experiment, we will see the flashes when electrons pass through the slits. However, under these experimental conditions, it becomes difficult to accurately determine the positions of the electrons or through which slit the electrons pass because the spatial resolution of the position decreases with increasing wavelength of the light. This type of experiment will produce a probability distribution curve of finding electrons on the wall as shown in Figure 1.4c. We will lose accuracy in recording both positions and interference of the electrons.

We should think about the meaning of this experiment. The experimental results show that detection of the electron positions at the slits disturbed the positions of electrons measured on the wall. The positions of electrons on the wall can be easily calculated from the *momentum* of incident electrons at the slits. Now, we have come to a very important conclusion that we cannot determine the position and momentum of electrons at the same time. In quantum mechanics, this is known as the *uncertainty principle*. If we try to determine the position of electrons very accurately, then we will lose all the information about their momentum, and the electrons are seen to behave like particles. On the other hand, if we try to determine the momentum of electrons very accurately, we will lose all the information about their position, and the electrons behave like waves. Thus, electrons behave as both particles and waves, and importantly, their behaviors change depending on the method we used to "observe" them. We may then ask, "What is the true character of electrons? Are they particles or waves?" The answer is we cannot define this with any great accuracy. There are limits to determining their precise nature, which are concepts described by the uncertainty principle in quantum mechanics. We do not know whether it is true or not but we must believe it because many modern electronic devices operate based on the theory of quantum mechanics and no contradiction to the uncertainty principle has been found even after nearly 100 years since the birth of quantum mechanics.

AN ASIDE: SCHRÖDINGER'S CAT

Schrödinger's cat is a thought experiment devised by Dr. Erwin Schrödinger. It attempts to illustrate what he saw as the problems of the interpretation of quantum mechanics. Schrödinger wrote

> A cat is penned in a steel chamber, along with the
> following device: in a Geiger counter there is a tiny
> bit of radioactive substance, so small, that perhaps

in the course of the hour one of the atoms decays, but also, with equal probability, perhaps none; if it happens, the counter tube discharges and through a relay releases a hammer which shatters a small flask of hydrocyanic acid. If one has left this entire system to itself for an hour, one would say that the cat lives if meanwhile no atom has decayed. The psi-function (wave function) of the entire system would express this by having in it the living and dead cat mixed or smeared out in equal parts.

In this thought experiment, Schrödinger pointed out that indeterminacy originally restricted to the atomic domain must be transformed into macroscopic indeterminacy, which could then be resolved by direct observation. Anyway, it is impossible for us to imagine the mixed state of the living and dead cat, although we may accept that the decay of atoms happens with a certain probability. How can we resolve this paradox? The answer according to Niels Bohr's group (Copenhagen interpretation) is as follows: a system stops being a superposition of states and becomes either one or the other when an observation takes place. Some interpret the experiment to mean that while the box is closed, the system simultaneously exists in a superposition of the states "decayed nucleus/dead cat" and "undecayed nucleus/living cat," and that only when the box is opened and an observation performed does the wave function collapse into one of the two states.

Albert Einstein did not agree with the concept of *probability* used in the Copenhagen interpretation. In a letter to Niels Bohr, he wrote the now famous words, "God does not throw dice." To which Niels Bohr wrote, "Einstein, do not tell God what to do."

There have been a lot of debates on the interpretation of quantum mechanics. Although the Copenhagen interpretation is currently accepted, it should be noted that the interpretation of quantum mechanics still has uncertainty.

1.1.3   Wave functions

In Equation 1.3, we defined the probability amplitude of finding an electron at a certain position. The term *amplitude* implies the existence of a wave. This wave is called as a *matter wave* and is represented by a function that includes the positions of particles and time. The function is known as a *wave function* represented by Ψ. The value of the wave function of a particle at a given point in space and time is related to the likelihood of the particle being there at that time. The probability of finding a particle in a volume element dv is proportional to $\Psi\Psi^*\,dv$, where * indicates the complex conjugate. If the wave function is normalized, the integral of the wave function over all space satisfies the following relation:

$$\int \Psi\Psi^*\,dv = \int |\Psi|^2\,dv = 1, \tag{1.4}$$

where $\Psi\Psi^*$ or $|\Psi^2|$ is referred to as the *probability density*. Notably, Equation 1.4 indicates that that $|\Psi^2|$ can be integrated over all space, which means the wave function becomes zero at infinity.

1.1.4   Operators

In quantum mechanics, physical quantities or actions employed to determine physical quantities (measurements) are represented by linear operators. Let us first see the property of operators. We define an operator as a rule by which, given any specific function, we can find another function. Thus, we may define an operator ξ as follows: multiply the function by the independent variable. This rule is written symbolically: $\xi f(x) = xf(x)$. Another operator ζ is the differentiation with respect to the independent variable: $\zeta f(x) = f'(x)$. The result obtained by the action of ξ followed by ζ is different from that obtained by the action of ζ followed by ξ as shown below:

$$\xi\zeta f(x) = \xi[\zeta f(x)] = xf'(x),$$
$$\zeta\xi f(x) = \zeta[\xi f(x)] = xf'(x) + f(x). \tag{1.5}$$

For two operators α and β, if $\alpha\beta$ and $\beta\alpha$ are the same, α and β are said to *commute* or be *commutable*. In this sense, the operators ξ and ζ are not commutable.

As mentioned in Section 1.1.2, a measurement changes the state of electrons represented by a wave function. Now assume we are going to measure physical quantities a and b of an electron by continuous measurements A and B. The result obtained by carrying out first measurement A followed by the measurement B is different from that obtained by conducting first measurement B followed by measurement A, because each of the measurements changes the wave function. This situation can be represented mathematically using operators acting on the wave function as follows:

$$AB\Psi \neq BA\Psi. \tag{1.6}$$

This equation indicates that the operators A and B are not commutable. In this case, measurements change the wave function and the two physical quantities corresponding to the operators A and B cannot be determined precisely at the same time due to laws defined by the uncertainty principle. On the other hand, if the operators A and B are commutable, we can determine the two physical constants at the same time without the restrictions of the uncertainty principle.

Operators corresponding to some important physical quantities are summarized as follows:

$$
\begin{aligned}
&\text{Position: } q_k && \text{operator: multiply } q_k, \\
&\text{Momentum: } p_k && \text{operator: } -i\hbar\,(\partial/\partial q_k') \\
&\text{Time: } t && \text{operator: multiply } t, \\
&\text{Energy: } E && \text{operator: } i\hbar\,(\partial/\partial t')
\end{aligned}
\tag{1.7}
$$

where $\hbar$ is given by $h/2\pi$ and $\hbar$ is Planck's constant.

Using the operators shown in Equation 1.7, we will proceed to consider the uncertainty principle for the determination of position and momentum of a particle. Now, Q and P are the operators corresponding to position and momentum, respectively. Let us calculate $(PQ - QP)\psi$. If it is not zero and has a certain value, then Q and P are not commutable and we cannot accurately determine the position and momentum of a particle at the same time. Since

$$
PQ\psi = -i\hbar\frac{\partial}{\partial q}(q\psi) = -i\hbar\psi - i\hbar q\frac{\partial\psi}{\partial q}
\tag{1.8}
$$

and

$$
QP\psi = q\left(-i\hbar\frac{\partial\psi}{\partial q}\right) = -i\hbar q\frac{\partial\psi}{\partial q},
\tag{1.9}
$$

we have the result that

$$
(PQ - QP)\psi = -i\hbar\psi \neq 0.
\tag{1.10}
$$

The rule of commutability and uncertainty principle was first proposed by Werner Heisenberg in 1927.

1.1.5 Eigenvalue and expected value

If the wave function Ψ is an *eigenfunction* of the operator A, we will have the following equation:

$$A\Psi = a\Psi, \tag{1.11}$$

where a is the *eigenvalue*. In this state, the value of constant a is measured precisely by the measurement corresponding to A. To obtain the real value of a, the operator A must be a *Hermitian operator* having the following property:

$$\int \psi^* A\psi \, dx = \int \psi A^* \psi^* \, dx. \tag{1.12}$$

In the case when the wave function is not an eigenfunction of an operator, Equation 1.11 is not valid, and the physical quantity obtained by measurement A scatters with repeated measurements. The expected value of the physical quantity $\langle A \rangle$ is given by

$$\langle A \rangle = \frac{\int \Psi^* A\Psi \, dv + \int \Psi A^* \Psi^* \, dv}{2\int \Psi^* \Psi \, dv}. \tag{1.13}$$

If the operator A is a Hermitian operator and the wave function is normalized as shown in Equation 1.4, then Equation 1.13 can be simplified to the form

$$\langle A \rangle = \int \Psi^* A\Psi \, dv. \tag{1.14}$$

1.1.6 Expansion theorem

If two functions $\varphi_1(x)$ and $\varphi_2(x)$ have the property that

$$\int \varphi_2^*(x) \, \varphi_1(x) \, dx = 0, \tag{1.15}$$

then they are said to be *orthogonal*. A set of functions

$$\varphi_1(x), \varphi_2(x), \varphi_3(x), \varphi_4(x), \ldots, \varphi_i(x) \tag{1.16}$$

such that any two functions in the set are orthogonal is called an *orthogonal set*. If in addition

$$\int \varphi_i^*(x) \, \varphi_i(x) \, dx = 1, \tag{1.17}$$

for all values of i, the set is called an *orthogonal and normalized system*.

The importance of orthogonal functions lies in the possibility of expanding arbitrary functions in a series of these orthogonal functions. If any function $f(x)$ can be expanded in a series,

$$f(x) = c_1\varphi_1(x) + c_2\varphi_2(x) + c_3\varphi_3(x) + c_4\varphi_4(x) + \cdots + c_i\varphi_i(x), \qquad (1.18)$$

using constant c_i and a set of functions $\varphi_i(x)$, the set of functions is called a *complete set*. If a complete set is an orthogonal and normalized system, the constant c_i can be determined by multiplying both sides of Equation 1.18 by $\varphi_i^*(x)$ as follows:

$$\int \varphi_i^*(x)f(x)\,dv = c_1 \int \varphi_i^*(x)\varphi_1(x)\,dv + \cdots + c_i \int \varphi_i^*(x)\varphi_i(x)\,dv + \cdots. \qquad (1.19)$$

By using the properties of an orthogonal and normalized system expressed by Equations 1.15 and 1.17, c_i is given by

$$c_i = \int \varphi_i^*(x)f(x)\,dv. \qquad (1.20)$$

1.1.7   Schrödinger equation

The Schrödinger equation formulated by Erwin Schrödinger in 1926 is an equation describing how a quantum state of a physical system varies. By specifying the total energy (Hamiltonian) of a quantum system, Schrödinger's equation can be solved, with the resulting solution being a wave function describing a quantum state.

It may be worthwhile to derive Schrödinger's equation because its derivation requires recalling the history of quantum mechanics. In 1900, Max Planck proposed that electromagnetic energy (radiation) could only be emitted in a quantized form; in other words, the energy could only be a multiple of an elementary unit according to the famous relationship

$$E = h\nu, \qquad (1.21)$$

where h is Planck's constant and ν is the frequency of the radiation. Albert Einstein's famous equation, proposed in his theory of special relativity in 1905, gives us an expression for the relationship between mass m and energy E of a particle

$$E = mc^2, \qquad (1.22)$$

where c is the velocity of light. In 1923, theoretician De Broglie stated that any moving particles or objects behaved like waves, which were called matter waves. Based on this theory, the two energies in Equations 1.21 and 1.22 are regarded as identical:

$$E = h\nu = mc^2 = pc, \tag{1.23}$$

where p is the momentum given by mc. The wave velocity c is equal to $\lambda\nu$, where λ is the wavelength, giving rise to the following relation:

$$E = h\nu = pc = p\lambda\nu, \quad \lambda = \frac{h}{p}. \tag{1.24}$$

Let us recall the seventeenth-century Newtonian mechanics, where the total energy of a particle is given by

$$E = \frac{1}{2}ms^2 + V = \frac{p^2}{2m} + V, \tag{1.25}$$

where s is the velocity of the particle and V is the potential energy. The solution for p from Equation 1.25 yields

$$p = \sqrt{2m(E-V)}. \tag{1.26}$$

From Equations 1.24 and 1.26, the particle wavelength is represented by

$$\lambda = \frac{h}{p} = \frac{h}{\sqrt{2m(E-V)}}. \tag{1.27}$$

The relationship between wavelength λ, wave velocity s, and frequency ν, that is, $s = \lambda\nu$, and Equation 1.27 yields

$$s^2 = \lambda^2\nu^2 = \frac{h^2\nu^2}{2m(E-V)}. \tag{1.28}$$

The general expression for the wave equation is

$$\frac{1}{v^2}\frac{\partial^2\Psi}{\partial t^2} = \Delta\Psi, \tag{1.29}$$

where $\Psi(x, y, z, t)$ is the wave function and Δ is Laplace's operator given by

$$\Delta = \frac{\partial^2}{\partial x^2} + \frac{\partial^2}{\partial y^2} + \frac{\partial^2}{\partial z^2}. \tag{1.30}$$

As a wave function Ψ, we assume the following formula:

$$\Psi(x, y, z, t) = \psi(x, y, z)\exp(\pm i\omega t), \quad \omega = 2\pi\nu. \tag{1.31}$$

By incorporating this wave function into Equation 1.29, we will have Schrödinger's equation

$$-\frac{2m(E-V)}{h^2\nu^2}\omega^2\psi(x) = \frac{\partial^2\psi(x)}{\partial x^2},$$

$$\left[-\frac{\hbar^2}{2m}\left(\frac{\partial^2}{\partial x^2} + \frac{\partial^2}{\partial y^2} + \frac{\partial^2}{\partial z^2}\right) + V\right]\psi(x) = E\psi(x), \tag{1.32}$$

$$H\psi = E\psi,$$

where H is the Hamiltonian operator,

$$H = -\frac{\hbar^2}{2m}\left(\frac{\partial^2}{\partial^2 x} + \frac{\partial^2}{\partial^2 y} + \frac{\partial^2}{\partial^2 z}\right) + V, \quad \hbar = \frac{h}{2\pi}. \tag{1.33}$$

Equation 1.32 is Schrödinger's equation for a stationary state, that is, for a state that is an eigenstate of the energy operator. The wave function in Equation 1.31 can be modified to

$$\Psi = \psi(x, y, z)\exp\left(\frac{-iEt}{\hbar}\right). \tag{1.34}$$

By using this formula, time-dependent Schrödinger's equation can be obtained as follows:

$$H\Psi = i\hbar\frac{\partial\Psi}{\partial t}. \tag{1.35}$$

1.1.8 Principle of superposition

If the wave functions Ψ_1 and Ψ_2 are solutions of one Schrödinger's equation, the linear combination of these functions, $c_1\Psi_1 + c_2\Psi_2$, is also the solution of Schrödinger's equation. This is called *the principle of superposition*.

Suppose that Ψ_1 corresponds to a state where electrons completely behave as particles and that Ψ_2 corresponds to a state where electrons completely behave as waves, the superimposed state of particles and waves represented by $c_1\Psi_1 + c_2\Psi_2$ is also considered to be a state satisfying the same Schrödinger's equation. In this way, the wave–particle duality, demonstrated in the double-slit experiments, is incorporated in the mathematical formulations of quantum mechanics.

If Ψ_1 and Ψ_2 are two eigenfunctions of the operator A, with eigenvalues of a_1 and a_2 $(a_1 \neq a_2)$, then the state represented by $c_1\Psi_1 + c_2\Psi_2$ is not an eigenfunction of A, since

$$A\Psi = A(c_1\Psi_1 + c_2\Psi_2) = c_1 a_1 \Psi_1 + c_2 a_2 \Psi_2 \neq a(c_1\Psi_1 + c_2\Psi_2). \qquad (1.36)$$

In the case when the wave functions belong to an orthogonal and normalized system, we obtain the relationship

$$\int (c_1^* \Psi_1^* + c_2^* \Psi_2^*)\,(c_1\Psi_1 + c_2\Psi_2)\,dv = c_1^* c_1 + c_2^* c_2 = 1. \qquad (1.37)$$

In a state represented by $c_1\Psi_1 + c_2\Psi_2$, the probability of observing the physical quantity corresponding to operator A to be a_1 is $c_1 c_1^*$ and the probability of observing the physical quantity to be a_2 is $c_2 c_2^*$.

1.1.9   Examples of solutions of the Schrödinger equation

As the derivations of the solutions of the Schrödinger equation are described in many textbooks on quantum mechanics, only the solutions will be presented here.

1.1.9.1   Electron in a one-dimensional (1D) box

We now consider an electron of mass m constrained to move in a fixed region of space, which for simplicity we take to be a 1D box with length a along the x-axis (Figure 1.5a). The potential energy is then

$$V = 0, \quad 0 < x < a,$$
$$V = \infty, \quad x \leq 0, \quad \text{or} \quad x \geq a. \qquad (1.38)$$

The solution of the Schrödinger equation under these boundary conditions gives the wave function and the energy levels as shown in Figure 1.5b and c:

$$\psi_n = \sqrt{\frac{2}{d}}\sin\frac{n\pi}{d}x, \quad E_n = \frac{n^2 h^2}{8ma^2}, \quad n = 1, 2, 3, \ldots. \qquad (1.39)$$

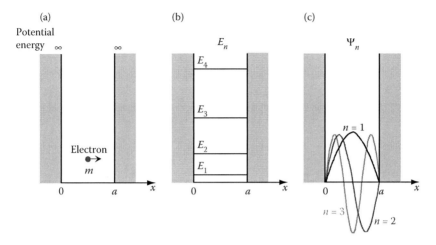

Figure 1.5 (See color insert following page 148.) An electron in a 1D box: (a) potential energy, (b) energy levels, and (c) wave functions.

1.1.9.2 Harmonic oscillator

Some quantum systems can be approximated by harmonic oscillators. For example, the variation of diatomic molecules and motions of atoms in a crystal lattice can be treated to a first approximation as the motion of particles in harmonic fields. The Schrödinger equation for the system in Figure 1.6a is given by

$$\frac{d^2\psi}{dx^2} + \frac{2m}{\hbar^2}\left(E - \frac{k}{2}x^2\right)\psi = 0, \tag{1.40}$$

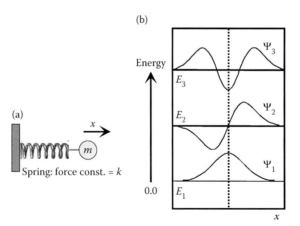

Figure 1.6 (a) Harmonic oscillator model and (b) the first few energy levels and the corresponding wave functions.

where k is the force constant. The solution of the Schrödinger equation yields the following wave functions and energies:

$$\psi_n = A_n H_n(z) \exp\left(-\frac{z^2}{2}\right) = A_n H_n \left[\left(\frac{mk}{\hbar^2}\right)^{1/4} x\right] \exp\left[-\frac{1}{2}\left(\frac{mk}{\hbar^2}\right)^{1/2} x^2\right],$$

(1.41)

$$E_n = \hbar \sqrt{\frac{k}{m}}\left(n + \frac{1}{2}\right) = \left(n + \frac{1}{2}\right)\hbar w_c, \quad n = 0, 1, 2, 3, \ldots,$$

where A_n and $H_n(z)$ are given by

$$A_n = \left[\frac{(mk/\hbar)^{1/4}}{2^n n! \pi^{1/2}}\right]^{1/2}, \quad H_n(z) = (-1)^n \exp(z^2)\frac{d^n}{dz^n}\exp(-z^2). \quad (1.42)$$

The first few energy levels and the corresponding wave functions are shown graphically in Figure 1.6b. We note that the wave functions are alternately symmetrical and antisymmetrical about the origin. It is important that the laws of quantum mechanics do not permit the harmonic oscillator to have zero energy, where the smallest allowed energy value is the zero-point energy $1/2 \cdot \hbar w_c$. This is in accordance with the uncertainty principle; if the oscillator had zero energy, it would have zero momentum and would also be located exactly at the position of minimum potential energy. The necessary uncertainties in position and momentum give rise to the zero-point energy.

1.9.1.3 Hydrogen atom

The hydrogen atom consists of a proton of charge $+e$ and mass M and an electron of charge $-e$ and mass m. The electron is located at a distance $r = (x^2 + y^2 + z^2)^{1/2}$ from the position of the proton. The potential energy of the electron,

$$V = -\frac{e^2}{4\pi\varepsilon_0 r}, \quad (1.43)$$

gives the following Schrödinger equation:

$$\left[-\frac{\hbar^2}{2\mu}\left(\frac{\partial^2}{\partial x^2} + \frac{\partial^2}{\partial y^2} + \frac{\partial^2}{\partial z^2}\right) - \frac{e^2}{4\pi\varepsilon_0 r}\right]\psi(x,y,x) = E\psi(x,y,z), \quad (1.44)$$

where μ is the reduced mass represented by $\mu = (1/M + 1/m)^{-1}$. The Schrödinger equation in rectangular coordinates can be expressed in spherical coordinates as

$$\frac{1}{r^2}\frac{\partial}{\partial r}\left(r^2\frac{d\psi}{dr}\right) + \frac{1}{r^2\sin^2\theta}\frac{\partial}{\partial\theta}\left(\sin\theta\frac{\partial\psi}{\partial\theta}\right) + \frac{1}{r^2\sin^2\theta}\frac{\partial^2\psi}{\partial\phi^2}$$

$$+ \frac{2\mu}{\hbar^2}[E - V(r)]\psi = 0. \tag{1.45}$$

The wave function is assumed to be represented by the formula

$$\psi = R(r)\Theta(\theta)\Phi(\phi). \tag{1.46}$$

By incorporating this wave function into the Schrödinger equation as given in Equation 1.44, $R(r)$ is expressed as follows:

$$R_{n,l}(r) = \left(\frac{2}{na_0}\right)^{3/2}\left[\frac{(n-l-1)!}{2n[(n+l)!]^3}\right]^{1/2}\left(\frac{2r}{na_0}\right)^l\exp\left(-\frac{r}{na_0}\right)L_{n+l}^{2l+1}\left(\frac{2r}{na_0}\right),$$

$$\tag{1.47}$$

$$a_0 = \frac{\hbar^2 4\pi\varepsilon_0}{\mu e^2},$$

where $L_q^p(r)$ is called the generalized Laguerre polynomial and is given by

$$L_q^p(\rho) = \sum_{k=0}^{q-p}(-1)^{k+1}\frac{(q!)^2\rho^k}{(q-p-k)!(p+k)!k!}. \tag{1.48}$$

$\Theta(\theta)$ is represented by

$$\Theta_{l,m_l}(\theta) = \left[\frac{2l+1}{2}\frac{(l-|m_l|)!}{(l+|m_l|)!}\right]^{1/2}P_l^{|m_l|}(\cos\theta), \tag{1.49}$$

where $P_l^{|m_l|}(\xi)$ is the associated Legendre polynomial and is given by

$$P_l^{|m_l|}(\xi) = \frac{1}{2^l l!}(1-\xi^2)^{|m_l|/2}\frac{d^{l+|m_l|}}{d\xi^{l+|m_l|}}(\xi^2-1)^l. \tag{1.50}$$

$\Phi(\phi)$ is represented by

$$\Phi_{m_l}(\phi) = (2\pi)^{-1/2} \exp(im_l\phi). \tag{1.51}$$

From Equations 1.47, 1.49, and 1.51, the wave function can be expressed as

$$\psi_{n,l,m_l} = Ar^l \exp\left(-\frac{r}{na_0}\right) L_{n+l}^{2l+1}\left(\frac{2r}{na_0}\right) P_l^{|m_l|}(\cos\theta) \exp(im_l\phi), \tag{1.52}$$

where A is a normalizing constant, and quantum parameters n, l, and m_l are given by

$$n = 1, 2, 3, \ldots,$$
$$l = 0, 1, 2, 3, \ldots, \tag{1.53}$$
$$ml = -l, -l + 1, \ldots, 0, \ldots, l - 1, l.$$

The quantum parameter n determines the energy of the atom as follows:

$$E_n = -\frac{\mu e^4}{2n^2\hbar^2}\frac{1}{(4\pi\varepsilon_0)^2}. \tag{1.54}$$

The quantum number l determines the total angular momentum and the quantum number m determines the z component of angular momentum. Figure 1.7 shows some of the lowest energy levels and the radial wave function $R_{n,l}(r)$ of hydrogen. It should be noted that the radial wave function has a nonzero value at $r = 0$ only for those states for which $l = 0$, that is, only for those states that have no angular momentum.

1.1.10   Matrix mechanics and bra–ket (Dirac) notation

Matrix mechanics is a formulation of quantum mechanics created by Werner Heisenberg, Max Planck, and Pascual Jordan in 1925. In matrix mechanics, a quantum state is represented by a quantum state vector and the physical quantities, such as position and momentum of a particle, are represented by matrices. The time evolution of the state is governed by Heisenberg's equation, which will be derived in the next section.

Paul Dirac's bra–ket notation [2] is a concise and convenient way of describing quantum states in matrix mechanics. He introduced and defined the symbol $|i\rangle$ to represent a quantum state i. This is called a ket or a ket vector. Dirac's notation uses a ket vector $|i\rangle$ to describe a quantum

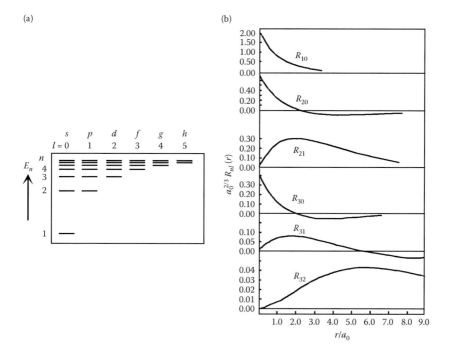

Figure 1.7 (a) Energy level and (b) radial eigenfunction $R_{nl}(r)$ of hydrogen.

state instead of a wave function ψ_i, which is the basis of Schrödinger's notation. The complex conjugate of the wave function ψ_i^* is represented by a bra vector $\langle i|$. Inner product of two complex functions can be described by the bra–ket notation as follows:

$$\int \psi_i^* \psi_j \, dv = \langle \psi_i | \psi_j \rangle = \langle i | j \rangle. \tag{1.55}$$

Some of the important rules explained in the previous sections will be summarized below using the bra–ket notation of matrix mechanics.

If a wave function is normalized, the following relation must be satisfied:

$$\int \psi \psi^* \, dv = \int |\psi|^2 \, dv = \langle \psi | \psi \rangle = 1. \tag{1.56}$$

A physical quantity or an action to determine physical quantities (measurements) is represented by a matrix **A** instead of a linear operator.

$$A\psi \rightarrow \mathbf{A} | \psi \rangle. \tag{1.57}$$

The non-commutative relation $AB \neq BA$ is described using two matrices. The expected value in measurement A is represented by

$$\langle A \rangle = \int \psi^* A \psi \, dv = \langle \psi | A | \psi \rangle. \tag{1.58}$$

A complete set means that any ket vectors $|f\rangle$ can be represented by a linear combination of the complete set of ket vectors,

$$|f\rangle = c_1 |\psi_1\rangle + c_2 |\psi_2\rangle + \cdots + c_i |\psi_i\rangle + \cdots, \tag{1.59}$$

but no element of the set can be represented as a linear combination of the others. A complete set of ket vectors is referred to as a *basis* with analogy to linear algebra where a basis is a linearly independent spanning set in a given vector space. As described in the previous sections, a measurement changes a quantum state, which can be represented in the matrix mechanics by the change of the ket vector $|\psi\rangle$ by applying a matrix $\mathbf{A}$. The change of a ket vector is due to the change of the coefficients in the linear combination of a basis. The expansion theorem in an orthogonal and normalized system is described as follows:

$$\langle \psi_i | f \rangle = c_1 \langle \psi_i | \psi_1 \rangle + \cdots + c_i \langle \psi_i | \psi_i \rangle \quad \therefore c_i = \langle \psi_i | f \rangle. \tag{1.60}$$

The Schrödinger equation for a stationary state is represented by

$$H_{ij} |\psi_j\rangle = E_i |\psi_i\rangle, \tag{1.61}$$

where H_{ij} is a Hamiltonian matrix. In matrix mechanics, a set of eigenvalues E_i can be obtained through the diagonalization of the Hamiltonian matrix instead of solving the Schrödinger equation. The principle of superposition is expressed as follows: if the two ket vectors $|\Psi_1\rangle$ and $|\Psi_j\rangle$ satisfy the Schrödinger equation above, then a new vector $c_1 |\Psi_1\rangle + c_2 |\Psi_2\rangle$ also satisfies the same Schrödinger equation. In this state, the probability of observing an energy to be E_1 is $c_1 c_1^*$ and the probability of observing the energy to be E_2 is $c_2 c_2^*$.

1.1.11 Comparison of the Heisenberg and Schrödinger approaches to quantum mechanics

It is worthwhile to compare some of the ideas of Heisenberg and Schrödinger while describing the concept of quantum mechanics. Let us recall the wave function in Equation 1.34 and the time-dependent Schrödinger equation in Equation 1.35. In the Schrödinger approach, which is sometimes called the *Schrödinger picture*, time is included in the wave function but the Hamiltonian operator is independent of time. The wave function, including its time evolution, is determined by the time-dependent Schrödinger equation. On the other hand, in the *Heisenberg picture*, time is involved in the matrix but the basis is independent of time. This situation will be explained using the Schrödinger picture. If a wave function at $t = 0$ is $\Psi(x, 0)$, the wave function at t is given by

$$\Psi(x, t) = \Psi(x, 0) \exp\left(-i\frac{E}{\hbar}t\right) \quad \text{(see Equation 1.34).} \tag{1.62}$$

The expected value of a measurement corresponding to an operator A is

$$\langle A \rangle = \int \Psi^*(x, t) A \Psi(x, t)\, dv$$

$$= \int \Psi^*(x, 0) \exp\left(i\frac{E}{\hbar}t\right) A \exp\left(-i\frac{E}{\hbar}t\right) \Psi(x, 0)\, dv. \tag{1.63}$$

In the Schrödinger picture, the two exponential terms are incorporated in the wave functions, whereas in the Heisenberg picture, they are incorporated in the Hermitian linear operator A such that

$$A(t) = \exp\left(i\frac{E}{\hbar}t\right) A \exp\left(-i\frac{E}{\hbar}t\right). \tag{1.64}$$

Let us derive the Heisenberg equation using matrix mechanics. The time-dependent Schrödinger equation can be written using Dirac's bra–ket notation as

$$H(t)\,|\,\Psi(t)\rangle = i\hbar\frac{d}{dt}\,|\,\Psi(t)\rangle, \tag{1.65}$$

where $H(t)$ is a time-dependent Hamiltonian matrix. The ket vector at t can be represented in terms of the ket vector at $t = 0$ as

$$|\,\Psi(t)\rangle = \exp\left(-i\frac{H}{\hbar}t\right)|\,\Psi(0)\rangle. \tag{1.66}$$

The expected value of a measurement corresponding to a matrix **A** is

$$\langle A \rangle = \langle \Psi(0)\,|\exp\left(i\frac{H}{\hbar}t\right) A \exp\left(-i\frac{H}{\hbar}t\right)|\,\Psi(0)\rangle. \tag{1.67}$$

Now we define a time-dependent matrix $\mathbf{A}(t)$ as

$$\mathbf{A}(t) = \exp\left(i\frac{H}{\hbar}t\right) A \exp\left(-i\frac{H}{\hbar}t\right). \tag{1.68}$$

This equation is identical to Equations 1.1 through 1.64. Assuming that A in the right-hand side of Equation 1.68 is independent of time, which means that classical time variation of observable A is not taken into account, differentiation of $\mathbf{A}(t)$ with respect to t yields

$$\frac{d}{dt}\mathbf{A}(t) = i\frac{1}{\hbar}H\exp\left(i\frac{H}{\hbar}t\right)A\exp\left(-i\frac{H}{\hbar}t\right) - i\frac{1}{\hbar}\exp\left(i\frac{H}{\hbar}\right)AH\exp\left(-i\frac{H}{\hbar}\right),$$

$$= i\frac{1}{\hbar}\{H\mathbf{A}(t) - \mathbf{A}(t)H\}, \tag{1.69}$$

and we will have the following Heisenberg equation:

$$\frac{d\mathbf{A}(t)}{dt} = i\frac{1}{\hbar}[H, \mathbf{A}(t)], \tag{1.70}$$

where $[A, B]$ means $AB - BA$. The Heisenberg equation represents the relationship of matrix elements between H and A. If the matrices $\mathbf{A}$ and $\mathbf{X}$ are commutable such that $AH - HA = 0$, we obtain

$$\frac{d}{dt}\mathbf{A}(t) = 0, \tag{1.71}$$

implying a steady state where the physical quantity corresponding to $\mathbf{A}$ is independent of time.

In the Schrödinger picture, wave functions describing quantum states vary with time, giving rise to the time evolution of physical quantities obtained by applying linear operators to the wave functions. On the other hand, in the Heisenberg picture, matrices corresponding to physical quantities vary with time, which is similar to the classical picture where physical quantities vary with time. In some sense, the Heisenberg picture is more natural and fundamental than the Schrödinger picture.

1.1.12 *Perturbation theory*

Quantum mechanics is only able to accurately solve a limited range of problems. Thus, most of the quantum mechanical problems are solved by approximate methods, and the most important method of approximations is the theory of quantum mechanical perturbation. Perturbation theory comprises mathematical methods that are used to find an approximate solution to a problem that cannot be solved exactly, by starting from the exact solution of a related problem.

Suppose that we wish to define the motion of a system whose Hamiltonian operator H is slightly different from the Hamiltonian operator H_0 of a problem that has already been solved. The Schrödinger equation for a problem described by H_0 is

$$H_0 \psi_n^0 = E_n^0 \psi_n^0, \quad n = 1, 2, 3, \ldots, \tag{1.72}$$

where wave functions with different values of n have a set of eigenfunctions with eigenvalues $E_1^0, E_2^0, E_3^0, \ldots$ If H is slightly different from H_0, we can write

$$H = H_0 + \lambda H^{(1)}, \tag{1.73}$$

where λ is a parameter and the term $\lambda H^{(1)}$, known as a *perturbation*, is small in comparison with H_0. We require a solution of the equation

$$H \psi_n = E \psi_n,$$

$$(H_0 + \lambda H^{(1)}) \psi_n = E_n \psi_n. \tag{1.74}$$

If λ has a small value, it would be natural to assume that the perturbation $\lambda H^{(1)}$ slightly changes the *unperturbed* eigenvalues ψ_n^0 and eigenfunctions E_n^0. Since ψ_n and E_n are functions of λ, we can expand them in the form of a power series

$$\psi_n = \psi_n^0 + \lambda \psi_n^{(1)} + \lambda^2 \psi_n^{(2)} + \cdots,$$

$$E_n = E_n^0 + \lambda E_n^{(1)} + \lambda^2 E_n^{(2)} + \cdots. \tag{1.75}$$

If we substitute these equations into Equation 1.74, we find

$$H_0 \psi_n^0 + \lambda(H^{(1)} \psi_n^0 + E_n^0 \psi_n^{(1)}) + \lambda^2(H^{(1)} \psi_n^{(1)} + H_0 \psi_n^{(2)}) + \cdots$$

$$= E_n^0 \psi_n^0 + \lambda(E_n^{(1)} \psi_n^0 + E_n^0 \psi_n^{(1)}) + \lambda^2(E_n^{(2)} \psi_n^0 + E_n^{(1)} \psi_n^{(1)} + E_n^0 \psi_n^{(2)}) + \cdots \tag{1.76}$$

In order for all these equations to be satisfied for all values of λ, the coefficients of the powers of λ on both sides of the equation must be equal. Equating the coefficients of the powers of λ yields the series of equations as follows:

$$H_0 \psi_n^0 = E_n^0 \psi_n^0, \tag{1.77}$$

$$(H_0 - E_n^0) \psi_n^{(1)} = E_n^{(1)} \psi_n^0 - H^{(1)} \psi_n^0, \tag{1.78}$$

$$(H_0 - E_n^0) \psi_n^{(2)} = E_n^{(2)} \psi_n^0 + E_n^{(1)} \psi_n^{(1)} - H^{(1)} \psi_n^{(1)}. \tag{1.79}$$

The first of these equations has already been solved. If the second can be solved, then we can determine $\psi_n^{(1)}$ and $E_n^{(1)}$. The solution of the third equation yields $\psi_n^{(2)}$ and $E_n^{(2)}$, and so on.

Let us solve Equation 1.78 by assuming that the function $\psi_n^{(1)}$ is expanded in terms of the normalized and orthogonal sets of functions, $\psi_1^0, \psi_2^0, \ldots, \psi_n^0, \ldots$, as

$$\psi_n^{(1)} = A_1 \psi_1^0 + A_2 \psi_2^0 + \cdots + A_m \psi_m^0 + \cdots, \tag{1.80}$$

where the coefficients A_m are to be determined. The function $H^{(1)} \psi_n^0$ can also be expanded in the series

$$H^{(1)} \psi_n^0 = H_{1n}^{(1)} \psi_1^0 + H_{2n}^{(1)} \psi_2^0 + \cdots + H_{mn}^{(1)} \psi_m^0 + \cdots, \tag{1.81}$$

where

$$H_{mn}^{(1)} = \int \psi_m^{0*} H^{(1)} \psi_n^0 \, dv. \tag{1.82}$$

Substituting this series into Equation 1.78 yields

$$(H_0 - E_n^0)(A_1 \psi_1^0 + A_2 \psi_2^0 + \cdots) = E_n^{(1)} \psi_n^0 - H_{1n}^{(1)} \psi_1^0 - H_{2n}^{(1)} \psi_2^0 - \cdots, \tag{1.83}$$

which can be reduced using Equation 1.77 to

$$(E_1^0 - E_n^0)A_1 \psi_1^0 + (E_2^0 - E_n^0)A_2 \psi_2^0 + \cdots = E_n^{(1)} \psi_n^0 - H_{1n}^{(1)} \psi_1^0 - H_{2n}^{(1)} \psi_2^0 \cdots. \tag{1.84}$$

If we pick up the term including ψ_n^0, we find

$$(E_n^0 - E_n^0)A_n \psi_n^0 = (E_n^{(1)} - H_{nn}^{(1)})\psi_n^0,$$

$$\therefore E_n^{(1)} - H_{nn}^{(1)} = 0, \quad E_n^{(1)} = H_{nn}^{(1)} = \int \psi_n^{0*} H^{(1)} \psi_n^0 \, dv. \tag{1.85}$$

By equating the coefficient of ψ_m^0 $(m \neq n)$ on both sides of Equation 1.84 to determine coefficients A_m, we obtain

$$(E_m^0 - E_n^0)A_m = -H_{mn}^{(1)}, \quad A_m = \frac{H_{mn}^{(1)}}{E_n^0 - E_m^0}. \tag{1.86}$$

This relationship gives us the values of all the A's except A_n. The coefficient A_n can be determined by the requirement for normalizing ψ_n. We can express ψ_n by substituting Equation 1.80 into Equation 1.75 as

$$\psi_n = \psi_n^0 + \lambda \sum_m{}' A_m \psi_m^0 + \lambda A_n \psi_n^0 + \lambda^2(\cdots),$$

$$\psi_n^* = \psi_n^{0*} + \lambda \sum_m{}' A_m \psi_m^{0*} + \lambda A_n \psi_n^{0*} + \lambda^2(\cdots),$$

(1.87)

where $\sum_m'$ means summation over all values of m except n. Then,

$$\int \psi^* \psi_n \, dv = 1 + 2\lambda A_n + \lambda^2(\cdots),$$

(1.88)

if the function ψ_n is normalized, then the right-hand side of this equation must be equal to unity for all values of λ, so that we must put A_n equal to zero, and the results to the first order in λ are

$$E_n = E_n^0 + \lambda H_{nn}^{(1)} + \lambda^2(\cdots),$$

$$\psi_n = \psi_n^0 + \lambda \sum_m{}' \frac{H_{mn}^{(1)}}{E_n^0 - E_m^0} \psi_m^0 + \lambda^2(\cdots).$$

(1.89)

By a similar analysis, we obtain expressions for E_n and ψ_n including second-order perturbation λ^2,

$$E_n = E_n^0 + \lambda H_{nn}^{(1)} + \lambda^2 \sum_m{}' \frac{H_{nm}^{(1)} H_{mn}^{(2)}}{E_n^0 - E_m^0} + \lambda^3(\cdots) + \cdots,$$

$$\psi_n = \psi_n^0 + \lambda \sum_m{}' \frac{H_{mn}^{(1)}}{E_n^0 - E_m^0} \psi_m^0 + \lambda^2 \sum_k{}'$$

(1.90)

$$\times \left(\sum_m{}' \frac{H_{km}^{(1)} H_{mn}^{(1)}}{(E_n^0 - E_k^0)(E_n^0 - E_m^0)} - \frac{H_{nn}^{(1)} H_{kn}^{(1)}}{(E_m^0 - E_k^0)^2} \right) \psi_k^0 + \lambda(\cdots).$$

As an example, let us derive an equation for the electronic polarizability of an atom by considering the perturbation of a weak electric field [3].

As given by Equation 1.75, the energy E_n is represented as

$$E_n = E_n^0 + \lambda E_n^{(1)} + \lambda^2 E_n^{(2)} + \cdots.$$

From Equations 1.89 and 1.90, energies $E_n^{(1)}$ and $E_n^{(2)}$ can be expressed using the bra–ket notation

$$E_0^{(1)} = \langle 0 | H^{(1)} | 0 \rangle, \tag{1.91}$$

$$E_0^{(2)} = -\sum_j \frac{\langle 0 | H^{(1)} | j \rangle \langle j | H^{(1)} | 0 \rangle}{E_j - E_0}. \tag{1.92}$$

To determine electronic polarization, we define the Hamiltonian operator as

$$H = H_0 - e \sum r_i \tilde{E}, \tag{1.93}$$

where e is the elementary electric charge, r_i is the displacement of electron i, and $\tilde{E}$ is the electric field.

The summation is taken over all electrons in the atom to give the dipole momentum induced by the electric field. $E_n^{(1)}$ is given by

$$E_0^{(1)} = \langle 0 | e \sum r_i | 0 \rangle \tilde{E}. \tag{1.94}$$

The sign of the operator $e \sum r_i$ changes according to the change of r_i to $-r_i$; therefore, integration over all atoms becomes zero, giving rise to $E_n^{(1)} = 0$. Then, $E_n^{(2)}$ is given by

$$E_0^{(2)} = -\sum_j \frac{\langle 0 | e \sum r_i | j \rangle \langle j | e \sum r_i | 0 \rangle}{E_j - E_0} \tilde{E}^2. \tag{1.95}$$

The change of energy due to the variation of polarization from P to $P + \delta P$ in a constant electric field E is $-E\delta P$, and we know the relationship $P = \sum \alpha$, where α is electric polarizability. The energy change from $P = 0$ to P is therefore given by

$$\Delta E = -\tfrac{1}{2} \alpha \tilde{E}^2, \tag{1.96}$$

where ΔE is the change in energy. Comparing Equations 1.95 and 1.96, we obtain

$$\alpha = 2 \sum_j \frac{\langle 0 | e \sum r_i | j \rangle \langle j | e \sum r_i | 0 \rangle}{E_j - E_0}. \tag{1.97}$$

Hartree's approximation is usually employed to calculate α. In this approximation, the atomic wave functions are written as one-electron wave functions, so that the above equation becomes

$$\alpha = 2 \sum_{n} \sum_{m} \frac{\langle \psi_n | er | \psi_m \rangle^2}{E_m - E_n},$$ (1.98)

where ψ_n and ψ_m are one-electron wave functions, $\sum_n$ is the summation over all occupied ground states, and $\sum_m$ is the summation over all excited states for which transitions are allowed. Equation 1.98 shows that the solids with a large energy gap $E_g = E_m - E_n$ have a high permittivity as shown below:

$$\text{Ge: } E_g = 0.7 \text{ eV}, \qquad \varepsilon_r = 16,$$

$$\text{Si: } E_g = 1.1 \text{ eV}, \qquad \varepsilon_r = 12,$$

$$\text{NaCl: } E_g = 9 \text{ eV}, \qquad \varepsilon_r = 2.25.$$

1.2 Electronic band structure of solids

The electronic state in solids is modified by the crystallographic periodicity. The electronic states in solids can be described in two ways: first, from the picture drawn for free electrons, which is called the *nearly free electron model*, and the second from the picture drawn for electrons bounded with atoms/ions, which is called the *tight binding model*. Either of these models will give us the same conclusion showing the formation of *energy bands* where many energy levels are distributed almost continuously as allowed energy levels for electrons and *energy gaps* where no energy level exists for electrons. An energy band and an energy gap are repeated alternately with increasing energy to form the electronic band structure of solids. In this section, we describe the concept of electronic band structure of solids (crystals). You can find more detailed explanations about this topic in many good textbooks [4].

1.2.1 Free electron Fermi gas

The solution of the Schrödinger equation for an electron in a 1D box was examined in Section 1.1.9. Electrons located in space with zero potential energy are called *free electrons*. An electron with mass m exists in a 1D box with length L. The potential energy is zero inside the box but is infinite out of the box as in Figure 1.5a. The Schrödinger equation for this electron can be written as

$$-\frac{\hbar^2}{2m} \frac{d^2 \psi_n}{dx^2} = E_n \psi_n.$$ (1.99)

By considering the boundary conditions, $\psi(0) = 0$ and $\psi(L) = 0$, the solution can be written as

$$\psi_n = A\sin\left(\frac{2\pi}{\lambda_n}x\right), \qquad \frac{1}{2}n\lambda_n = L, \qquad (1.100)$$

where n is an integer and A is a constant. A solution of the Schrödinger equation for the one-electron system is termed an *orbital*. The energy is given by

$$E_n = \frac{\hbar^2}{2m}\left(\frac{n\pi}{L}\right)^2. \qquad (1.101)$$

Let us put N electrons in the 1D box. In the one-electron approximation, the potential energy due to electron–electron interactions can be ignored in the Schrödinger equation, thus yielding the same wave function and energy as shown in Equations 1.100 and 1.101, respectively. The Pauli exclusion principle formulated by Wolfgang Pauli in 1925 dictates that no two electrons can have the same four quantum numbers, that is, if n, l, and m_l are the same, the electrons must have opposite spins. The spin quantum number, m_s, has values of $\pm 1/2$ corresponding to up and down spins. Particles that obey the Pauli exclusion principle are called *fermions* and a collection of noninteracting fermions is called a *Fermi gas* or *free electron gas*. Each orbital with an energy determined by the quantum number n in Equation 1.101 can be occupied by two electrons with opposite spins. If different physical states (orbitals) have the same energy level, the states are said to be *degenerate*. Now assume that we fill up the orbitals one by one with two electrons from low to high energy levels. The highest energy level occupied by electrons is called the *Fermi energy* E_F.

Let us expand the model to three dimensions, with an electron located in a three-dimensional (3D) cube with the edge L. The Schrödinger equation for this electron is given by

$$-\frac{\hbar^2}{2m}\left(\frac{\partial^2}{\partial^2 x} + \frac{\partial^2}{\partial^2 y} + \frac{\partial^2}{\partial^2 z}\right)\psi(r) = E\psi(r), \qquad (1.102)$$

where r is the position vector of the electron. The solution of this equation will yield the wave functions

$$\psi(r) = A\sin\left(\frac{\pi n_x x}{L}\right)\sin\left(\frac{\pi n_y y}{L}\right)\sin\left(\frac{\pi n_z z}{L}\right), \qquad (1.103)$$

where n_x, n_y, and n_z are integers. These wave functions can be generalized in the form of plane waves

$$\psi(r) = \exp(ik \cdot r), \tag{1.104}$$

where r is a *wavevector* that will be very important to describe the electronic structure of solids.

Now suppose that the cube is repeated three-dimensionally to give a periodicity of wave functions as follows:

$$\psi(x + L, y, z) = \psi(x, y, z), \tag{1.105}$$

then substitute Equation 1.105 for the wave function in Equation 1.104. The x component of the wave function should satisfy the following condition:

$$\exp[ik_x(x + L)] = \exp(ik_x x)\exp(ik_z L) = \exp(ik_x x) \tag{1.106}$$

and k_x should be

$$k_x = 0, \pm\frac{2\pi}{L}, \pm\frac{4\pi}{L}, \dots \tag{1.107}$$

The same condition results for k_y and k_z. By substituting Equation 1.104 for the wave function into Equation 1.102, we obtain

$$E_k = \frac{\hbar^2}{2m}\lfloor k \rfloor^2 = \frac{\hbar^2}{2m}(k_x^2 + k_y^2 + k_z^2). \tag{1.108}$$

Let us recall the operator representing momentum (Equation 1.7)

$$-i\hbar\frac{\partial}{\partial q_k} = -i\hbar\left(\frac{\partial}{\partial x}, \frac{\partial}{\partial y}, \frac{\partial}{\partial z}\right) = -i\hbar\nabla. \tag{1.109}$$

By applying this operator to the wave function in Equation 1.104, we obtain

$$-i\hbar\nabla\psi_k(r) = \hbar\vec{k}\psi_k(r). \tag{1.110}$$

This equation implies that the wave functions of plane waves are eigenfunctions of the operator representing momentum and the eigenvalues are given by $\hbar(k_x, k_y, k_z)$, indicating that the wavevector is proportional to the momentum vector.

Now we define a unit reciprocal vector $a^* = (a_x^*, a_y^*, a_z^*)$, where a_x^*, a_y^* and a_z^* are unity but their dimensions are reciprocal to length. Reciprocal space is defined by a collection of vectors given by $a^* = (c_1 a_x^*, c_2 a_y^*, c_3 a_z^*)$, where c_1, c_2, and c_3 are arbitrary real numbers. The wavevectors k satisfying Equation 1.107 are reciprocal vectors. Since these k vectors are discrete, a collection of k will make a 3D lattice in the reciprocal space, which is called a *reciprocal lattice*. If N (an even number) electrons are located in the cube, then these electrons occupy energy levels given by Equation 1.108 from low to high energy levels. The energy levels and orbitals occupied by the electrons having k-values (k_x, k_y, k_z) are represented by a sphere in the reciprocal space as shown in Figure 1.8. The sphere surface is called a *Fermi surface* because the energy at the surface of the sphere is the Fermi energy E_F. Now, if we define the k-values on the surface as k_F, we obtain the following relationship:

$$E_F = \frac{\hbar^2}{2m} k_F^2. \tag{1.111}$$

From the condition given in Equation 1.107, a single k-value (k_x, k_y, k_z) exists per volume element $(2\pi/L)^3$ in the reciprocal space. Therefore, the number of orbitals in a sphere with a volume of $4\pi k_F^3/3$ is given by

$$\frac{4\pi k_F^3/3}{(2\pi/L)^3} = \frac{V}{6\pi^2} k_F^3 = \frac{N}{2}, \tag{1.112}$$

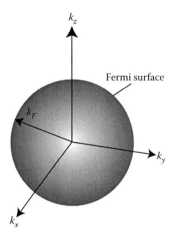

Figure 1.8 Fermi surface in the reciprocal space.

where V ($=1/L^3$) is the volume of the reciprocal unit cell. From this equation, the value of k_F is given by

$$k_F = \left(\frac{3\pi^2 N}{V} \right)^{1/3}.$$ (1.113)

By substituting the above relationship into Equation 1.111, we obtain

$$E_F = \frac{\hbar^2}{2m} \left(\frac{3\pi^2 N}{V} \right)^{1/3}.$$ (1.114)

We will now formulate the number of states per unit energy dN/dE. Assuming an arbitral energy E, the above equation can be rewritten as

$$N = \frac{V}{3\pi^2} \left(\frac{2mE}{\hbar^2} \right)^{3/2},$$ (1.115)

$$D(E) \equiv \frac{dN}{dE} = \frac{V}{2\pi^2} \left(\frac{2m}{\hbar^2} \right)^{3/2} E^{1/2},$$ (1.116)

where $D(E)$ is the *density of states (DOS)*.

1.2.2 Nearly free electron model (DOS)

The electronic band structure of crystalline solids can be described by the nearly free electron model, where free electrons are perturbed by a potential formed by a periodic arrangement of ions in a crystalline material. The important feature of waves traveling in a crystal is Bragg's diffraction. Suppose that electrons are located in a 1D periodic potential with an interatomic distance L as shown in Figure 1.9a. If Bragg's diffraction takes place at the line "a" in Figure 1.9a, the diffraction angle of $2\theta = 180°$ gives the condition for Bragg's diffraction, $2L \sin \theta = n\lambda$, as follows:

$$k = \frac{2\pi}{\lambda} = \frac{n\pi}{L}, \quad n = \pm1, \pm2, \pm3, \ldots.$$ (1.117)

The wave functions at $n = \pm1$ are not identical to the case of free electrons given by $\exp(+i\pi x/L)$ and $\exp(-i\pi x/L)$, but the wave propagating along the +x-direction and that along the −x-direction interact to form a

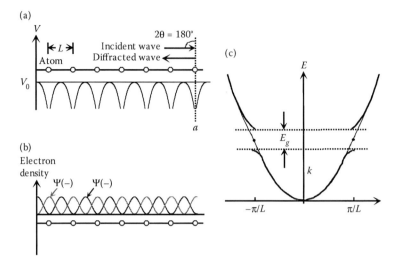

Figure 1.9 Bragg's diffraction and wave function in a periodic potential and formation of the energy gap.

standing wave. The standing wave is represented by the following two formulas:

$$\psi(+) = \exp\left(i\frac{\pi x}{L}\right) + \exp\left(-i\frac{\pi x}{L}\right) = 2\cos\left(\frac{\pi x}{L}\right),$$

$$\psi(-) = \exp\left(i\frac{\pi x}{L}\right) - \exp\left(-i\frac{\pi x}{L}\right) = 2i\sin\left(\frac{\pi x}{L}\right).$$

(1.118)

The probability density of electrons shown in Figure 1.9b indicates that the $|\Psi(+)|^2$ has a minimum, while $|\Psi(-)|^2$ has a maximum at the ionic position, giving rise to the different potential energies of electrons. This implies that the energy of the orbitals allowed at $k = \pm\pi/\lambda$ has two values at the same time. The difference in energy produces an energy gap E_g at $k = \pm\pi/\lambda$ as shown in Figure 1.9c, although the E versus k curve is continuous in the case of free electrons. If the potential energy is approximated to be $V(x) = V_0\cos(2\pi x/L)$, where $V_0 > 0$, the energy gap is given by

$$E_g = \int_0^L V(x)\left[|\psi(-)|^2 - |\psi(+)|^2\right]dx = \frac{2V_0}{L}\int_0^L \cos\left(\frac{2\pi x}{L}\right)$$

$$\times\left[\sin^2\left(\frac{\pi x}{L}\right) - \cos^2\left(\frac{\pi x}{L}\right)\right]dx = V_0.$$

(1.119)

This relationship indicates that E_g is equal to the Fourier component of the potential energy.

1.2.3 Bloch function

The Bloch function is the wave function of a particle (usually, an electron) placed in a periodic potential. It is given by

$$\psi(k) = \exp(ik \cdot r) \cdot u(r). \tag{1.120}$$

The Bloch function consists of the product of a plane wave envelope function and a periodic function $u(r)$, which has the same periodicity as the potential. The result that the eigenfunctions can be written in this form for a periodic system is called Bloch's theorem. When $u(r)$ is constant, the Bloch function becomes $\psi(k) = \exp(ik \cdot r)$, which is identical to the wave function of free electrons (Equation 1.104).

1.2.4 Krönig–Penny model

The Krönig–Penny model is a simple model for describing the behavior of electrons in a periodic crystal structure exhibiting band structure. The Krönig–Penny potential is shown in Figure 1.10a [4]. The potential is $V(x) = 0$, $0 < x < a$ and $V(x) = V_0$, $a \leq x < (a + b)$. The lattice spacing is $(a + b)$ and the potential repeats itself periodically with this period.

Schrödinger's equation to be solved in this case is

$$-\frac{\hbar^2}{2m}\frac{d^2\psi}{dx^2} + V(x)\psi = E\psi. \tag{1.121}$$

In the region, $0 < x < a$, the wave function is represented by the sum of plane waves propagating on the left- and right-hand sides as follows:

$$\psi = A\exp(ikx) + B\exp(-ikx) \tag{1.122}$$

(a)   (b)

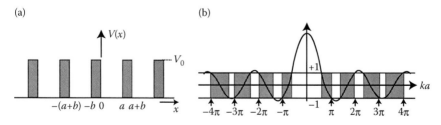

Figure 1.10 (a) Potential used in the Krönig–Penny model. (b) Change of (p/ka) $\sin ka + \cos ka$ with ka when $P = 3\pi/2$. This function should not exceed ±1 as indicated in Equation 1.130.

and the energy is given by (see Equation 1.108)

$$E = \frac{\hbar^2 k^2}{2m}. \tag{1.123}$$

In the region, $a \leqq x < (a + b)$, the wave function is

$$\psi = C\exp(Qx) + D\exp(-Qx), \tag{1.124}$$

where Q satisfies the following relationship:

$$V_0 - E = \frac{\hbar^2 Q^2}{2m}. \tag{1.125}$$

The coefficients A, B, C, and D must be determined so as to satisfy the boundary conditions, where ψ and $d\psi/dx$ are continuous at $x = 0$. In this case, we obtain

$$\psi: A + B = C + D, \tag{1.126}$$

$$\frac{d\psi}{dx}: ikA - ikB = QC - QD. \tag{1.127}$$

The same boundary conditions at $x = a$ yield

$$\psi: A\exp(ika) + B\exp(-ika) = C\exp(Qa) + D\exp(-Qa), \tag{1.128}$$

$$\frac{d\psi}{dx}: ikA\exp(ika) - ikB\exp(-ika) = QC\exp(Qa) - QD\exp(-Qa). \tag{1.129}$$

Since Equations 1.126 through 1.129 are homogeneous equations with coefficients A, B, C, and D, the conditions for obtaining solutions are given by solving secular equations. Then, we will obtain

$$\left[\frac{Q^2 - k^2}{2Qk}\right]\sinh Qb \sin ka + \cosh Qb \cos ka = \cos k(a+b). \tag{1.130}$$

The above equation can be simplified by taking the limits $b \to 0$, $V_0 \to \infty$ without changing $V_0 b$,

$$\left(\frac{P}{ka}\right)\sin ka + \cos ka = \cos ka. \tag{1.131}$$

The left-hand side of the above equation with $P = 3\pi/2$ is shown as a function of ka in Figure 1.10b. Since $-1 \leq \cos ka \leq 1$, the left-hand side must not exceed ±1 to satisfy the above equation. The range of ka to have a solution is shown in Figure 1.10b. The allowed k-values are limited in some ranges, which means that the allowed energies are also limited in some ranges because energy is a function of k: $k = (2mE/\hbar^2)^{1/2}$ from Equation 1.123. The gray parts in Figure 1.10b form the allowed energy bands and the other parts form the forbidden bands. Figure 1.11a shows the relationship between the energy E and the wave number k. The region $-\pi/a \le k \le \pi/a$ is known as the first Brillouin zone, the regions $-2\pi/a \le k \le \pi/a$ and $\pi/a \le k \le 2\pi/a$ are referred to as the second Brillouin zone, which continues to the third, fourth, and so on. The wave function is continuous in the Brillouin zone and the energy gap E_g is formed at the zone boundary. The E versus k relationship can be reduced within the first Brillouin zone as shown in Figure 1.11b, which is called the reduced-zone scheme. The E versus k relationship drawn by the reduced-zone scheme is called the energy band diagram.

1.2.5 Tight binding model

The wave function of an electron in a crystal is represented by the Bloch function, $\psi(k) = \exp(ik \cdot r) \cdot u(r)$ (Equation 1.120). If the $u(r)$ is constant, the electron is regarded as a free electron. On the other hand, if the $\exp(ik \cdot r)$ is constant, the electron is completely bound with an ion having a periodic arrangement in the crystal. A real crystal must have an intermediate state and approaches from both sides give the same answer. In the tight binding model, it is assumed that an electron strongly bound with an ion centered at a lattice point moves in the crystal under the influence of other ions.

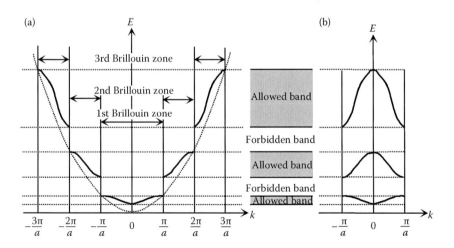

Figure 1.11 (a) Energy band diagram obtained for the Krönig–Penny model. (b) Reduced-zone scheme of the energy band diagram.

The Hamiltonian operator of an electron bound with an isolated ion at a point R_l is defined as

$$H_l = -\frac{\hbar^2}{2m}\Delta + U(r - R_l),\qquad(1.132)$$

where $U(r - R_l)$ is the potential of the isolated ion. The Schrödinger equation is

$$H_l\phi(r - R_l) = E_0\phi(r - R_l),\qquad(1.133)$$

where $\phi(r - R_l)$ is the localized wave function (atomic orbital). Then, the Bloch function $\psi_k(r)$ of an electron in a crystal is given by

$$\psi_k(r) = \sum_l \exp(i k \cdot R_l)\phi(r - R_l).\qquad(1.134)$$

The Hamiltonian operator of the electron in a crystal is assumed to be

$$H = -\frac{\hbar^2}{2m}\Delta + V(r).\qquad(1.135)$$

By using the Hamiltonian operator in Equation 1.132, the above Hamiltonian operator is

$$H = -\frac{\hbar^2}{2m}\Delta + V(r) = H_l + V(r) - U(r - R).\qquad(1.136)$$

The $H - H_l = V - U$ will be treated as a perturbation (see Section 1.1.12). Then, the energy is

$$E = \frac{\int \psi_k^*(r) H \psi_k(r)\, dr}{\int \psi_k^*(r)\psi_k(r)\, dr} \cong E_0 + \frac{1}{N}\int \psi_k^* \sum_l \exp(i k \cdot R_l)(H - H_l)\phi(r - R)\, dr,$$

$$(1.137)$$

where the overlapping of atomic orbitals is ignored and the denominator is assumed to be N, which is the number of ions in the crystal. By substituting Equation 1.134 into the above equation, we obtain

$$E = E_0 + \frac{1}{N}\sum_m \sum_l \exp\left[i k(R_m - R_l)\right]\int \phi^*(r - R_l)(H - H_m)\phi(r - R_m)\, dr.$$

$$(1.138)$$

Since all lattice points are identical, R_m becomes the origin and ρ_l is newly defined to be $\rho_l = R_l - R_m$. The above equation becomes

$$E = E_0 + \sum_l \exp(-ik\rho_l) \int \phi^*(r - \rho_l)\{V(r) - U(r)\}\phi(r)dr. \tag{1.139}$$

The integral in the above equation can be divided into the following two parts:

$$\int \phi^*(r)\{V(r) - U(r)\}\phi(r)dr = -\alpha,$$

$$\int \phi^*(r - \rho_l)\{V(r) - U(r)\}\phi(r)dr = -\gamma. \tag{1.140}$$

By substituting these relationships into Equation 1.139, the energy is given by

$$E(k) = E_0 - \alpha - \gamma \sum_l \exp(-ik \cdot \rho_l). \tag{1.141}$$

If the crystal is a simple cube with lattice constant a, ρ_l will be

$$\rho_l = (\pm a, 0, 0), \ (\pm a, 0, 0), \ (\pm a, 0, 0), \ (\pm a, 0, 0), \tag{1.142}$$

and the energy is represented by

$$E(k) = E_0 - \alpha - 2\gamma(\cos k_x a + \cos k_y a + \cos k_z a). \tag{1.143}$$

The change of energy levels with the interionic distance is shown in Figure 1.12. With decreasing interionic distance, the degenerated, discrete energy levels of electrons in an ion spread out to form energy bands due to overlapping of atomic orbitals. From Equation 1.143, we know that the energy of the band center is lower than the atomic energy level by α and that the band width is 12γ. If the two bands do not overlap at an interionic distance a, then an energy gap is formed as in the case of the nearly free electron model.

1.2.6 Phase velocity, group velocity, and effective mass

We have to understand the concepts of phase velocity, group velocity, and effective mass of electrons in a crystal to know the motion of electrons in a crystal, which is different from those in a free space.

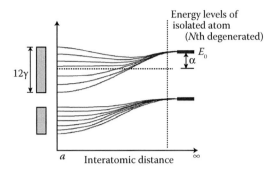

Figure 1.12 Formation of energy bands using the tight binding model.

Now, assume that we are looking at waves of the sea in front of us as they are propagating from the left- to the right-hand side. In order to measure the velocity of a specific wave, we have to move in the same direction as the wave, increasing our moving velocity. When the wave looks stationary, our velocity is the same as that of the phase velocity of the wave. If we define the wave by the equation, $A \cos(kx - \omega t)$, where A is a constant, k is a wave number, x is the direction of wave propagation, ω is an angular frequency, and t is the time, then the following condition must be satisfied at this phase velocity because the wave looks stationary:

$$kx - \omega t = \text{constant.} \tag{1.144}$$

By differentiating both sides of the above equation, we will obtain

$$\frac{dk}{dt}x + k\frac{dx}{dt} - \frac{d\omega}{dt}t - \omega = 0. \tag{1.145}$$

If we assume that x and k are independent of time, the above equation becomes

$$k\frac{dx}{dt} - \omega = 0, \qquad \frac{dx}{dt} = v_p = \frac{\omega}{k}, \tag{1.146}$$

where v_p is the phase velocity. The phase velocity can also be defined as

$$\omega = 2\pi f, \quad k = \frac{2\pi}{\lambda}, \quad v_p = f\lambda \rightarrow v_p = \frac{\omega}{k}, \tag{1.147}$$

where λ is the wavelength.

If the phase velocity is independent of frequency, a linear relationship between ω and k is obtained as shown in Figure 1.13a. However, an

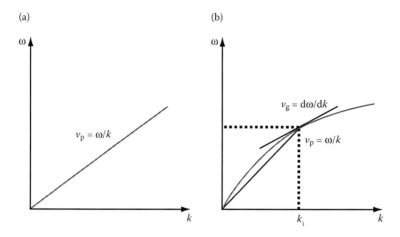

Figure 1.13 (a) A linear ω versus k relationship and (b) a nonlinear ω versus k relationship.

electron wave in a crystal does not meet this condition and the ω versus k relationship becomes nonlinear as shown in Figure 1.13b. If we focus on a wave with a wave number of k_1, we can define the velocity of this wave as

$$v_g = \frac{d\omega}{dk}, \tag{1.148}$$

where v_g is the group velocity.

It should be noted that angular frequency ω is proportional to energy E according to the relationship, $E = \hbar\omega$. The group velocity is given by

$$v_g = \frac{1}{\hbar}\frac{dE(k)}{dk}. \tag{1.149}$$

Let us now consider the equation of motion of an electron in a crystal under an electric field $\tilde{E}$. The work done by the electric field on the electron during time δt is

$$\delta E = -e\tilde{E}v_g\delta t, \tag{1.150}$$

which is equal to the change of energy,

$$\delta E = \left(\frac{dE}{dk}\right)\delta k = \hbar v_g\delta k, \tag{1.151}$$

then we obtain

$$\delta k = -\left(\frac{e\tilde{E}}{\hbar}\right)\delta t \quad \therefore \hbar\frac{dk}{dt} = -e\tilde{E} = F, \tag{1.152}$$

where F is the force applied to the electron. From the relationship $\hbar k = p$, where p is the momentum, we will know that Equation 1.152 corresponds to Newton's equation of motion. By differentiating the group velocity given in Equation 1.149, we obtain

$$\frac{dv_g}{dt} = \frac{1}{\hbar}\frac{d^2E}{dk\,dt} = \frac{1}{\hbar}\left(\frac{d^2E}{dk^2}\frac{dk}{dt}\right) = \left(\frac{1}{\hbar^2}\frac{d^2E}{dk^2}\right)F. \tag{1.153}$$

Then, we can define the mass

$$\frac{1}{m^*} = \frac{1}{\hbar^2}\frac{d^2E}{dk^2}, \tag{1.154}$$

where m^* is known as the *effective mass*. For a free electron, the E versus k relationship is quadratic, and thus the effective mass is constant and equal to the real mass of an electron. In a crystal, however, the situation is far more complex and the E versus k relationship is not even approximately quadratic in a wide range of k. However, wherever a minimum occurs in the relationship, the minimum can be approximated by a quadratic curve in a limited region of k around that minimum. Hence, for electrons that have energy close to a minimum, effective mass is a useful concept. On the other hand, for energy regions far away from a minimum, the effective mass can be negative or can even approach infinity. The effective mass is a tensor because it depends on the direction with respect to the crystal axes.

1.2.7 Reciprocal lattice and the Brillouin zone

Let us draw a 3D energy band diagram of a real crystal. All lattice point position vectors R in a crystal with a 3D periodicity are given by

$$R = n_1a_1 + n_2a_2 + n_3a_3, \tag{1.155}$$

where n_1, n_2, and n_3 are integers and $(a_1\ a_2\ a_3)$ are primitive vectors in the real space. Now, we will define the reciprocal lattice as

$$\frac{a_1^*}{2\pi} = \frac{a_2 \times a_3}{V_a}, \quad \frac{a_2^*}{2\pi} = \frac{a_3 \times a_1}{V_a}, \quad \text{and} \quad \frac{a_3^*}{2\pi} = \frac{a_1 \times a_2}{V_a}, \tag{1.156}$$

where $(a_1\, a_2\, a_3)$ are primitive vectors in real space and $a_1 \cdot (a_2 \times a_3)$ is the unit cell volume. The reciprocal lattice position vectors G are given by

$$G = m_1 a_1^* + m_2 a_2^* + m_3 a_3^*, \tag{1.157}$$

where m_1, m_2, and m_3 are integers. The lattice point position vectors (lattice vector) R have dimensions of [length], while that of the reciprocal lattice vector G is [length]$^{-1}$.

The relationship between the real lattice and the reciprocal lattice is as follows:

Real lattice   Reciprocal lattice

Simple lattice ⟷ Simple lattice

Body-centered lattice ⟷ Face-centered lattice

Face-centered lattice ⟷ Body-centered lattice

Figure 1.14 shows the transformation of lattice shapes from the real to reciprocal lattices.

The Wigner–Seitz cell is a kind of primitive cell used to represent the periodical arrangement of atoms in a crystal. The Wigner–Seitz cell around a lattice point is defined as the locus of points in space, which are closer to that lattice point than to any of the other lattice points. The construction of the two-dimensional (2D) Wigner–Seitz cell is done as follows: the cell may be chosen by first picking a lattice point and then drawing lines to all nearby (closest) lattice points. At the midpoint of each line, another line is drawn

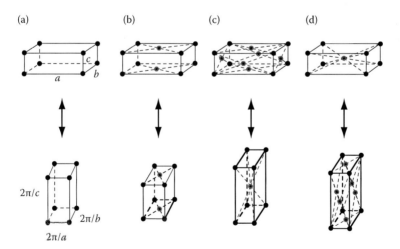

Figure 1.14 **(See color insert following page 148.)** (a) Simple lattice, (b) cubic-centered lattice, (c) face-centered lattice, and (d) body-centered lattice.

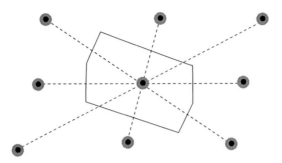

Figure 1.15 Example of 2D Wigner–Seitz cells.

normal to each of the first set of lines. An example of a 2D Wigner–Seitz cell is shown in Figure 1.15. In the case of a 3D lattice, a perpendicular plane is drawn at the midpoint of the lines between the lattice points. The smallest volume enclosed in this way is the Wigner–Seitz cell. All space within the lattice is filled by this type of primitive cell and there will not be any gaps. Examples of 3D Wigner–Seitz cells are shown in Figure 1.16.

The first Brillouin zone is defined as the Wigner–Seitz cell in reciprocal space (k-space). In the case of an electron in a 1D box of length L, the first Brillouin zone is in the region $-\pi/L < x < \pi/L$ and the second Brillouin zone is in the region $-2\pi/L < x < -\pi/L$, $\pi/L < x < 2\pi/L$. The wave function is continuous in the Brillouin zone and the energy gap is formed at the zone boundary as shown in Figure 1.9c. The first Brillouin zones of a body-centered cubic lattice and a face-centered cubic lattice in a real space are shown in Figure 1.17. The shape of the Brillouin zones should be compared with the Wigner–Seitz cell in Figure 1.16. A description of the several points of high symmetry that are of special interest (these are called critical points) in the Brillouin zone is given in Table 1.1.

(a)   (b)

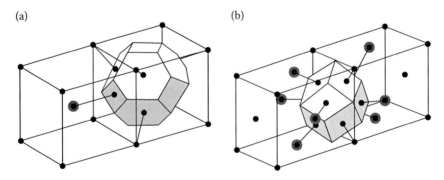

Figure 1.16 **(See color insert following page 148.)** (a) The Wigner–Seitz cell of a body-centered cubic (bcc) structure and (b) the Wigner–Seitz cell of a face-centered cubic (fcc) structure.

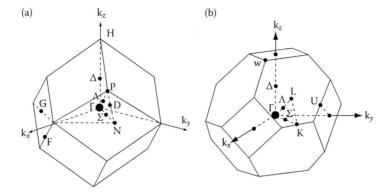

Figure 1.17 The first Brillouin zone of (a) a body-centered cubic lattice and (b) a face-centered cubic lattice in a real space.

Table 1.1 Description of Several Points of High Symmetry of Special Interest

Symbol	Description
Γ	Center of the Brillouin zone
Simple cube	
M	Center of an edge
R	Corner point
X	Center of a face
Face-centered cubic	
K	Middle of an edge joining two hexagonal faces
L	Center of a hexagonal face
U	Middle of an edge joining a hexagonal and a square face
W	Corner point
X	Center of a square face
Body-centered cubic	
H	Corner point joining four edges
N	Center of a face
P	Corner point joining three edges
Hexagonal	
A	Center of a hexagonal face
H	Corner point
K	Middle of an edge joining two rectangular faces
L	Middle of an edge joining a hexagonal and a rectangular face
M	Center of a rectangular face

1.2.8   *Energy band structure of silicon (Si)*

As an example, the energy band structure of Si is shown in Figure 1.18. The energy band below the forbidden band is the valence band, which is filled with outer electrons in the Si atoms. The valence band derived from the 3p orbitals along the x-, y-, and z-directions has triply degenerated states at $k = 0$ (Γ-point). If an electron starts to move along the x-direction (one of Δ-axis, $k \neq 0$), the triply degenerated states split into one single state (x-direction) and the doubly degenerated states (y- and z-directions) as shown in 1 and 2 in Figure 1.18. Similarly, the triply degenerated states split into three different states on the Σ-axis. The k-dependence of energy is different along the x-, y-, and z-axes, giving rise to a nonspherical Fermi surface. The energy band above the forbidden band is the conduction band. The energy difference between the valence and the conduction bands is ~3 eV at the Γ-point, ~4 eV at the X-point, and ~3 eV at the L-point. The minimum energy difference is 1.17 eV, which is lower than these values and is known as the energy gap, E_g. The characteristic feature of the energy band diagram of Si is

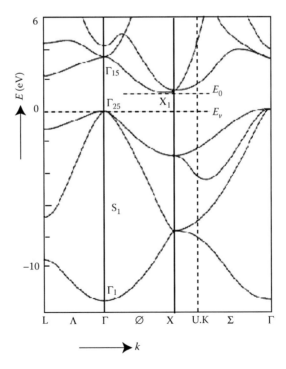

Figure 1.18 Energy band diagram of silicon (Si).

that the energy gap is not located at the Γ-point but at $ka/\pi \sim 0.85$, away from the Γ-point.

1.2.9 Tight binding approximation for calculating the band structure of graphene

The band diagram of a graphite sheet (known as graphene) is calculated using the tight binding approximation.

The arrangement of carbon atoms in a graphite sheet is illustrated in Figure 1.19. The atomic arrangement is honeycomb-like, and the unit cell is described by 2D unit vectors, a_1 and a_2, as shown in the figure. One unit cell contains two independent atomic sites, A- and B-sites. Vectors denoted by t_1, t_2, and t_3 are those from the A-sites, adjacent to the B-sites. Similarly, vectors from the B-site adjacent to A-sites are described by $-t_1$, $-t_2$, and $-t_3$.

In the tight binding approximation, assuming that carbon atoms interact with their outermost core's electron orbital $\phi(r)$, then the wave function of the valence electron is described as

$$\psi_k(r) = C_A \sum_{R_A} e^{ik \cdot R_A} \phi(r - R_A) + C_B \sum_{R_B} e^{ik \cdot R_B} \phi(r - R_B), \tag{1.158}$$

where R_A and R_B represent the position of the A-site and B-site atoms, and C_A and C_B are the coefficients of the linear combination. The terms $e^{ik \cdot R_A}$ and $e^{ik \cdot R_B}$ are derived from Bloch's theorem. The wave function given by Equation 1.158 is the solution of the following Schrödinger equation:

$$H\psi_k(r) = \varepsilon(k)\psi_k(r). \tag{1.159}$$

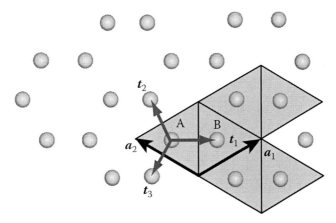

Figure 1.19 **(See color insert following page 148.)** Arrangement of atoms in a graphite layer.

Substituting Equation 1.158 into Equation 1.159, we obtain

$$C_A \sum_{R_A} e^{ik \cdot R_A} H\phi(r - R_A) + C_B \sum_{R_B} e^{ik \cdot R_B} H\phi(r - R_B)$$

$$= C_A \varepsilon(k) \sum_{R_A} e^{ik \cdot R_A} \phi(r - R_A) + C_B \varepsilon(k) \sum_{R_B} e^{ik \cdot R_B} \phi(r - R_B). \tag{1.160}$$

Operating $\phi^*(r - R_{AA})$ from the left-hand side of the above equation followed by integration of the equation over all real space, we obtain

$$\langle \phi(r - R_{AA}) | H | \psi_k(r) \rangle$$

$$= C_A \sum_{R_A} e^{ik \cdot R_A} \langle \phi(r - R_{AA}) | H | \phi(r - R_A) \rangle + C_B \sum_{R_B} e^{ik \cdot R_B} \langle \phi(r - R_{AA}) | H | \phi(r - R_B) \rangle$$

$$\left\{
\begin{aligned}
& \langle \phi(r - R_{AA}) | H | \phi(r - R_A) \rangle = \alpha \delta_{A,AA} \\
& \langle \phi(r - R_{BB}) | H | \phi(r - R_B) \rangle = \alpha' \delta_{B,BB} \\
& \langle \phi(r - R_{AA}) | H | \phi(r - R_B) \rangle = \beta \quad \text{AA \& B: the nearest neighbor} \\
& \qquad\qquad\qquad\qquad\qquad\quad = 0 \quad \text{others} \\
& \langle \phi(r - R_{BB}) | H | \phi(r - R_A) \rangle = \beta \quad \text{BB \& A: the nearest neighbor} \\
& \qquad\qquad\qquad\qquad\qquad\quad = 0 \quad \text{others} \\
& \langle \phi(r - R_2) | \phi(r - R_1) \rangle = \delta_{1,2}
\end{aligned}
\right. \tag{1.161}$$

$$= C_A e^{ik \cdot R_{AA}} \alpha + C_B \left\{ \exp(ik \cdot (R_{AA} + t_1))\beta + \exp(ik \cdot (R_{AA} + t_2))\beta \right.$$

$$\left. + \exp(ik \cdot (R_{AA} + t_3))\beta \right\}$$

$$= C_A e^{ik \cdot R_{AA}} \alpha + C_B e^{ik \cdot R_{AA}} \Gamma(k)\beta$$

$$= \langle \phi(r - R_{AA}) | \varepsilon(k) | \psi_k(r) \rangle = C_A \varepsilon(k) e^{ik \cdot R_{AA}}.$$

After all,

$$C_A(\alpha - \varepsilon(k)) + C_B \Gamma(k)\beta = 0, \tag{1.162}$$

where $\Gamma(k)$ is given by

$$\Gamma(k) = \exp(ik \cdot t_1) + \exp(ik \cdot t_2) + \exp(ik \cdot t_3). \tag{1.163}$$

In the same way, operating $\phi^*(r - R_{BB})$ from the left-hand side of Equation 1.160 followed by integration of the equation over all space: that is, $\langle \phi(r - R_{BB}) | H | \psi_k(r) \rangle = \langle \phi(r - R_{BB}) | \varepsilon(k) | \psi_k(r) \rangle$, we obtain

$$C_A \Gamma^*(k)\beta + C_B(\alpha' - \varepsilon(k)) = 0. \tag{1.164}$$

To keep Equations 1.162 and 1.164 meaningful for nontrivial coefficients C_A and C_B, the following condition is necessary:

$$\det \begin{bmatrix} \alpha - \varepsilon(k) & \Gamma(k)\beta \\ \Gamma^*(k)\beta & \alpha' - \varepsilon(k) \end{bmatrix} = 0. \tag{1.165}$$

Solving this equation, we finally obtain the band dispersion as follows:

$$\varepsilon(k) = \frac{\alpha + \alpha'}{2} \pm \sqrt{\left(\frac{\alpha - \alpha'}{2}\right)^2 + \beta^2 |\Gamma(k)|^2}. \tag{1.166}$$

Both the A- and B-sites are occupied by carbon atoms in graphite, which means $\alpha = \alpha'$. Therefore, the above dispersion relation can be simplified as

$$\varepsilon(k) = \alpha \pm \beta |\Gamma(k)|. \tag{1.167}$$

Information about the Brillouin zone is necessary for a graphical representation of the band structure. In the 2D x–y coordinates illustrated in Figure 1.20, the vectors a_1, a_2, t_1, t_2, and t_3 are described by two components as follows:

$$a_1 = (a, 0),$$

$$a_2 = \left(-\frac{a}{2}, \frac{\sqrt{3}}{2}a\right), \tag{1.168}$$

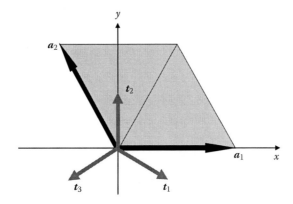

Figure 1.20 **(See color insert following page 148.)** Unit cell of a graphite layer.

$$t_1 = \frac{a}{\sqrt{3}}\left(\frac{\sqrt{3}}{2}, -\frac{1}{2}\right),$$

$$t_2 = \frac{a}{\sqrt{3}}(0, 1),$$

$$t_3 = \frac{a}{\sqrt{3}}\left(-\frac{\sqrt{3}}{2}, -\frac{1}{2}\right). \tag{1.169}$$

The 2D reciprocal vectors are given by

$$b_1 = \frac{2\pi}{a}\left(1, \frac{1}{\sqrt{3}}\right),$$

$$b_2 = \frac{2\pi}{a}\left(0, \frac{2}{\sqrt{3}}\right). \tag{1.170}$$

The Brillouin zone is defined by the area enclosed by the lines that bisect the reciprocal vectors b_1, b_2, $-b_1 + b_2$, $-b_1$, $-b_2$, and $b_1 - b_2$ as illustrated in Figure 1.21.

The band diagram of the k-dependence of electronic energy is given by

$$\varepsilon(k) = \alpha \pm \beta |\Gamma(k)|. \tag{1.171}$$

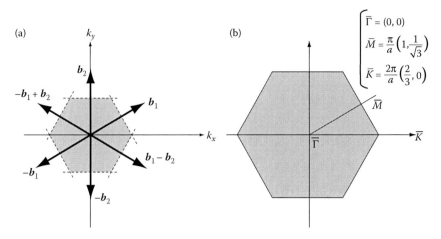

Figure 1.21 (a) Reciprocal lattice vectors and (b) the first Brillouin zone of a graphite layer.

By substituting Equations 1.163 and 1.169 into the above equation, we obtain the mathematical expression for the band dispersion of graphene as follows:

$$
\begin{aligned}
\varepsilon(k) &= \alpha \pm \beta \left| \exp\left\{ i\frac{a}{\sqrt{3}}\left(\frac{\sqrt{3}}{2}k_x - \frac{1}{2}k_y \right) \right\} + \exp\left\{ i\frac{a}{\sqrt{3}}k_y \right\} \right. \\
&\quad \left. + \exp\left\{ i\frac{a}{\sqrt{3}}\left(-\frac{\sqrt{3}}{2}k_x - \frac{1}{2}k_y \right) \right\} \right|, \\
&= \alpha \pm \beta \left| \exp\left\{ i\left(\frac{a}{2}k_x - \frac{a}{2\sqrt{3}}k_y \right) \right\} + \exp\left\{ i\frac{a}{\sqrt{3}}k_y \right\} \right. \\
&\quad \left. + \exp\left\{ i\left(-\frac{a}{2}k_x - \frac{a}{2\sqrt{3}}k_y \right) \right\} \right|.
\end{aligned}
$$

(1.172)

The electronic energies at $\bar{\Gamma}$: $(k_x = 0, k_y = 0)$, $\bar{M}$: $(k_x = \pi/a, k_y = \pi/\sqrt{3}a)$, and $\bar{K}$: $(k_x = 4\pi/3a, k_y = 0)$ in the Brillouin zone are given by

$$
\varepsilon(\bar{\Gamma}) = \alpha \pm 3\beta
$$

$$\varepsilon(\bar{M}) = \alpha \pm \beta \left| \exp\left\{ i\left(\frac{a}{2}\frac{\pi}{a} - \frac{a}{2\sqrt{3}}\frac{\pi}{\sqrt{3}a} \right) \right\} + \exp\left\{ i\frac{a}{\sqrt{3}}\frac{\pi}{\sqrt{3}a} \right\} \right.$$

$$\left. + \exp\left\{ i\left(-\frac{a}{2}\frac{\pi}{a} - \frac{a}{2\sqrt{3}}\frac{\pi}{\sqrt{3}a} \right) \right\} \right|,$$

$$= \alpha \pm \beta \left| \exp\left\{ i\frac{\pi}{3} \right\} + \exp\left\{ i\frac{\pi}{3} \right\} + \exp\left\{ -i\frac{2\pi}{3} \right\} \right|,$$

$$= \alpha \pm \beta \left| \frac{1}{2} + \frac{\sqrt{3}}{2} i \right|,$$

$$= \alpha \pm \beta,$$

$$\varepsilon(\bar{K}) = \alpha \pm \beta \left| \exp\left\{ i\left(\frac{a}{2}\frac{4\pi}{3a} \right) \right\} + \exp\{0\} + \exp\left\{ i\left(-\frac{a}{2}\frac{4\pi}{3a} \right) \right\} \right|,$$

$$= \alpha \pm \beta \left| \exp\left\{ i\frac{2\pi}{3} \right\} + 1 + \exp\left\{ -i\frac{2\pi}{3} \right\} \right|,$$

$$= \alpha \pm \beta \cdot 0,$$

$$= \alpha. \tag{1.173}$$

The energy band structure of graphene obtained by connecting $\varepsilon(k)$ plots is illustrated in Figure 1.22.

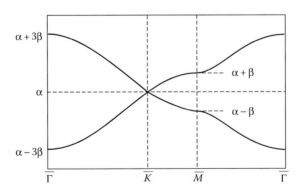

Figure 1.22 Band dispersion of a graphite layer.

1.2.10 Electron correlation

In the previous sections, we neglected the electron–electron interactions (electron correlations) in the calculations of electronic states. However, the electron correlations sometimes give rise to important effects in the properties of solids. In the first part of this section, a mean field approximation where the effect of electron correlations is approximated as an effective potential from surrounding electrons is described. The Hartree–Fock approximation is one of the simplest approaches in the mean field approximations. A more powerful tool to calculate the electronic state including electron–electron interactions is the density functional method, which will be discussed in the later part.

1.2.10.1 Hartree–Fock approximation

We assume that the Hamiltonian operator of a system with N electrons is given by the sum of the single-electron operator H_i including the kinetic energy and Coulomb interaction with the nucleus, and the Coulomb repulsion e^2/r_{ij} between two electrons by

$$H = \sum_{i=1}^{N} H_i + \sum_{i<j} \frac{e^2}{r_{ij}}. \tag{1.174}$$

The simplest way to express the wave function is as a product of single-electron states. However, there are two important requirements that must be taken into account. One is that the wave functions have the spin coordinate as well as the spatial coordinate because an electron has the spin degree of freedom. To meet this requirement, the single-electron state is expressed as $\Psi_{i,\sigma}(\tau_i) = \Psi_{i,\sigma}(r_i, \xi_i) = \Psi_i(r_i)\chi_\sigma(\xi_i)$, where $\Psi_i(r_i)$ is the orbital function, r_i is the position of the ith electron, $\chi_\sigma(\xi_i)$ is the spin function, and ξ_i is the spin coordinate. The other requirement is the antisymmetrization of the N-electron wave function because an electron is a Fermion. Such an N-electron wave function can be written in the form of a Slater determinant as a first approximation,

$$\Phi(\tau_1, \tau_2 \ldots, \tau_N) = \frac{1}{\sqrt{N!}} \begin{vmatrix} \Psi_{1\sigma_1}(\tau_1) & \Psi_{2\sigma_2}(\tau_1) & \cdots & \Psi_{N\sigma_N}(\tau_1) \\ \Psi_{1\sigma_1}(\tau_2) & \Psi_{2\sigma_2}(\tau_2) & \cdots & \Psi_{N\sigma_N}(\tau_2) \\ \vdots & \vdots & \vdots & \vdots \\ \Psi_{1\sigma_1}(\tau_N) & \Psi_{2\sigma_2}(\tau_N) & \cdots & \Psi_{N\sigma_N}(\tau_N) \end{vmatrix}, \tag{1.175}$$

where the single-electron states $\Psi_{i,\sigma}(\tau_i)$ make a complete orthonormal set. At the lowest energy state within this approximation, the

N-electron wave function should minimize an expectation value of the energy E

$$E \equiv \int \cdots \int \Phi^* H \Phi \, d\tau_1 \, d\tau_2 \cdots d\tau_N,$$

$$= \sum_{i=1}^{N} \langle \Psi_{i\sigma_i} | H | \Psi_{i\sigma_i} \rangle$$

$$+ \frac{1}{2} \sum_{i,j=1}^{N} \left\{ \langle \Psi_{i\sigma_i} \Psi_{j\sigma_j} | \frac{e^2}{r_{ij}} | \Psi_{i\sigma_i} \Psi_{j\sigma_j} \rangle - \langle \Psi_{i\sigma_i} \Psi_{j\sigma_j} | \frac{e^2}{r_{ij}} | \Psi_{j\sigma_j} \Psi_{i\sigma_i} \rangle \right\}, \quad (1.176)$$

where $\langle \Psi_i \Psi_j | e^2/r_{ij} | \Psi_i \Psi_j \rangle$ and $\langle \Psi_i \Psi_j | e^2/r_{ij} | \Psi_j \Psi_i \rangle$ are the Coulomb integral and exchange integral, respectively. Therefore, using the variation method with respect to $\Psi_i^*(\tau_i)$, the following closed equation is obtained:

$$\left[H(r_1) + \sum_{j \neq i} \int \Psi_{j\sigma_j}^*(r_2) \frac{e^2}{r_{12}} \Psi_{j\sigma_j}(r_2) \, dr_2 \right] \Psi_{i\sigma_i}(r_1)$$

$$- \sum_{j \neq i, \sigma_j} \delta_{\sigma_i \sigma_j} \left[\int \Psi_{j\sigma_j}^*(r_2) \frac{e^2}{r_{12}} \Psi_{i\sigma_i}(r_2) \, dr_2 \right] \Psi_{j\sigma_j}(r_1) = \varepsilon_i \Psi_{i\sigma_i}(r_1). \quad (1.177)$$

The second term on the left-hand side of Equation 1.177 is the exchange potential term as a consequence of the Fermi–Dirac statistics. Since the exchange potential is nonlocal, Equation 1.177 has to be solved self-consistently. The exchange potential term can be rewritten as follows:

$$V_{ex}^{i\sigma}(r_1) \Psi_{i\sigma}(r_1) = \left[\int \rho_{ex}^{i\sigma}(r_1, r_2) \frac{e^2}{r_{12}} \, dr_2 \right] \Psi_{i\sigma}(r_1), \quad (1.178)$$

$$\rho_{ex}^{i\sigma}(r_1, r_2) = - \frac{\Psi_{i\sigma}^*(r_1) \Psi_{i\sigma}(r_2) \sum_j \Psi_{j\sigma}^*(r_2) \Psi_{j\sigma}(r_1)}{|\Psi_{i\sigma}(r_1)|^2} \quad (1.179)$$

where $V_{ex}^{i\sigma}$ is the exchange potential for $\Psi_{i\sigma}(r_1)$. For simplicity, we take an average of $V_{ex}^{i\sigma}$ using the weight $W^{i\sigma}(r_1)$:

$$W^{i\sigma}(r_1) = \frac{|\Psi_{i\sigma}(r_1)|^2}{\sum_i |\Psi_{i\sigma}(r_1)|^2}. \quad (1.180)$$

Then, the exchange potential can be written as

$$V_{ex}^{\sigma}(r_1) = \sum_i W^{i\sigma}(r_1) V_{ex}^{i\sigma}(r_1) = \int dr_2 n_{ex}^{\sigma}(r_1, r_2) \frac{e^2}{r_{12}} \qquad (1.181)$$

and the single-electron density $n^s(r_1)$ and the exchange hole density $n_{ex}^{\sigma}(r_1, r_2)$ are given as

$$n^{\sigma}(r_1) = \sum_i |\Psi_{i\sigma}(r_1)|^2 \qquad (1.182)$$

$$n_{ex}^{\sigma}(r_1, r_2) = -\left| \sum_i \Psi_{i\sigma}^*(r_1) \Psi_{i\sigma}(r_2) \right|^2 / n^{\sigma}(r_1). \qquad (1.183)$$

If all the wave functions are approximated as plane waves,

$$\Psi_k(r) = \frac{1}{\sqrt{V}} e^{ik \cdot r}. \qquad (1.184)$$

Equation 1.181 is reduced to

$$V_{ex}^{\sigma}(r) = \int dr_2 \left[-\frac{9}{2} n \left\{ \frac{j_1(k_F r_{12})}{k_F r_{12}} \right\}^2 \right] \frac{e^2}{r_{12}} = -\frac{6}{\pi} e^2 k_F \int_0^{\infty} \frac{1}{x} \{j_1(x)\}^2 \, dx$$

$$= -\frac{3e^2}{2\pi} k_F = -3e^2 \left\{ \frac{3}{4\pi} n^{\sigma}(r) \right\}^{1/3}. \qquad (1.185)$$

It should be noted that the nonlocal exchange potential is reduced to the local potential as given by Equation 1.185, indicating that the exchange potential can be determined when the electron density is given as a function of position. The potential V_{ex}^{σ} is called the Slater's exchange potential.

Using this exchange potential, the Hartree–Fock equation in Equation 1.177 can be solved, giving the electronic states of solids. In order to take into account the electron correlation effects, Equation 1.185 should be replaced with Equation 1.186.

$$V_{x\alpha}^{\sigma}(r) = -3\alpha e^2 \left\{ \frac{3}{4\pi} n^{\sigma}(r) \right\}^{1/3}. \qquad (1.186)$$

This approach is called the X_α method, in which correlation effects are included in the adjustable parameter α [5].

1.2.10.2 Density functional method

The density functional method is the most popular approach for calculating the band structure at present. In this method, the single-electron density $n(r)$ and the total energy $E[n]$ can be expressed as

$$E[n] = \int dr\, v_{ext}(r)n(r) + F[n], \qquad (1.187)$$

$$F[n] = \langle \Psi \,|\, T + V_{ee} \,|\, \Psi \rangle, \qquad (1.188)$$

where v_{ext} is the external potential including the potential from the nucleus, T is the kinetic energy, V_{ee} is the electron–electron interaction, and Ψ is the wave function minimizing the expectation value of $T + V_{ee}$. In particular, the ground-state energy E_{GS} is given by the single-electron density of the ground state n_{GS},

$$E_{GS} = \int dr\, v_{ext}(r)n_{GS}(r) + F[n_{GS}]. \qquad (1.189)$$

In order to calculate the electronic state, we divide $F[n]$ into the following terms:

$$F[n] = T_0[n] + \frac{e^2}{2} \iint dr\, dr' \frac{n(r)n(r')}{|r - r'|} + E_{xc}[n] \qquad (1.190)$$

The first term on the right-hand side of Equation 1.190 is the kinetic energy of the ground state of a virtual noninteracting system, the second term is the Coulomb interaction between electrons, and the third term is the exchange and correlation energy. Let us assume that the virtual noninteracting system gives the density $n(r)$ in an effective single-electron potential $v_{eff}(r)$. In such a system,

$$\left\{ -\frac{\hbar^2}{2m}\Delta + v_{eff}(r) \right\} \Psi_i(r) = \varepsilon_i \Psi_i(r), \qquad (1.191)$$

$$n(r) = \sum_i |\Psi_i(r)|^2. \qquad (1.192)$$

Since $T_0[n]$ is written as

$$T_0[n] = \sum_i \varepsilon_i - \int d\mathbf{r} v_{\text{eff}}(\mathbf{r}) n(\mathbf{r}), \tag{1.193}$$

Equation 1.190 is represented as

$$E[n] = \sum_i \varepsilon_i - \int d\mathbf{r} v_{\text{eff}}(\mathbf{r}) n(\mathbf{r}) + \int d\mathbf{r} v_{\text{ext}}(\mathbf{r}) n(\mathbf{r})$$

$$+ \frac{e^2}{2} \iint d\mathbf{r} d\mathbf{r}' \frac{n(\mathbf{r}) n(\mathbf{r}')}{|\mathbf{r} - \mathbf{r}'|} + E_{\text{xc}}[n]. \tag{1.194}$$

To minimize $E[n]$, the electron density has to satisfy the variational equation,

$$\delta E[n] = \int d\mathbf{r}\, \delta n(\mathbf{r}) \left\{ \int d\mathbf{r}'' \frac{\delta v_{\text{eff}}(\mathbf{r}'')}{\delta n(\mathbf{r})} n(\mathbf{r}'') - v_{\text{eff}}(\mathbf{r}) - \int d\mathbf{r}'' \frac{\delta v_{\text{eff}}(\mathbf{r}'')}{\delta n(\mathbf{r})} n(\mathbf{r}'') \right.$$

$$\left. + v_{\text{ext}}(\mathbf{r}) + e^2 \int d\mathbf{r}' \frac{n(\mathbf{r}')}{|\mathbf{r} - \mathbf{r}'|} + \frac{\delta E_{\text{exc}}[n]}{\delta n(\mathbf{r})} \right\} = 0, \tag{1.195}$$

under the conservation condition of total electron number

$$\int d\mathbf{r}\, \delta n(\mathbf{r}) = 0. \tag{1.196}$$

From this equation, the potential $v_{\text{eff}}(\mathbf{r})$ can be expressed as

$$v_{\text{eff}}(\mathbf{r}) = v_{\text{ext}}(\mathbf{r}) + e^2 \int d\mathbf{r}' \frac{n(\mathbf{r}')}{|\mathbf{r} - \mathbf{r}'|} + \frac{\delta E_{\text{xc}}[n]}{\delta n(\mathbf{r})}. \tag{1.197}$$

The last term of Equation 1.197 is the exchange correlation potential and Equations 1.191, 1.192, and 1.197 are called Kohn–Sham equations [6].

To proceed with the calculation, we assume that the spatial variation of the density is not very steep and that the density can be approximated as the local value of a nearly uniform electron gas. This approximation is called local density approximation, where the electron correlation energy E_{xc} is written as

$$E_{\text{xc}}[n] = \int d\mathbf{r} \varepsilon_{\text{xc}}(n(\mathbf{r})) n(\mathbf{r}) \tag{1.198}$$

From Equation 1.197, the exchange potential $v_{xc}(r)$ is also written as

$$v_{xc}(r) = \frac{\delta E_{xc}[n]}{\delta n(r)} = \varepsilon_{xc}(n(r)) + n(r)\frac{d\varepsilon_{xc}(n)}{dn} \qquad (1.199)$$

Using an appropriate density $n(r)$, we can calculate the potential $v_{eff}(r)$ using Equations 1.197 and 1.199. The resulting $v_{eff}(r)$ allows us to solve Equations 1.191 and 1.192, giving $\Psi_i(r)$ and ε_i. Since the electronic states are occupied up to the Fermi energy, $n(r)$ can be calculated by summing Equation 1.192 up to the Fermi level. Repeating these procedures, we can finally obtain the electronic structure of this system. More detailed issues are beyond the scope of this textbook but further details can be found in Refs. [7–9].

1.3 Material properties with respect to characteristic size in nanostructures

The physical properties of materials depend on their size because phenomena determining materials characteristics undergo qualitative and/or quantitative changes as the size of physical objects decreases. The most dramatic physical changes occur on scales where the quantum nature of objects starts dominating their properties, that is, on scales of 0.1–1 nm, even though long-range electromagnetic interactions in the regions 10–100 nm can be an important factor for many properties.

The next three chapters deal with the electrical, optical, and magnetic properties of materials as functions of size and distance. Chapter 2 describes electronic states in nanomaterials, where the reduction in size to the nanometer scale significantly increases the surface–volume ratio and electrostatic capacity to enhance the charging and surface effects, which gives rise to the quantum size effects. Next, Chapter 3 focusses on changes of the optical properties and interaction of light with matter that occurs as the size of material objects decreases to nanometer scales. Some of the most dramatic changes in optical properties of materials occur on the scales of 0.1–10 nm. Finally, Chapter 4 focusses on the fundamental physical descriptions of the magnetic and magnetotransport properties of nanostructures, including the Kubo effect, novel magnetic ordering of transition metal surfaces, superparamagnetism, and domain wall effects.

Figure 1.23 depicts the typical length scales of nanostructures in materials and physical phenomena as the origins of electrical, optical, and magnetic properties. As shown in Figure 1.23a, the basic building blocks of materials are atoms with dimensions of the order of the Bohr radius (0.05 nm; the radius of hydrogen atom). Solids are regarded as a condensed state of these atoms with appropriate interatomic distances.

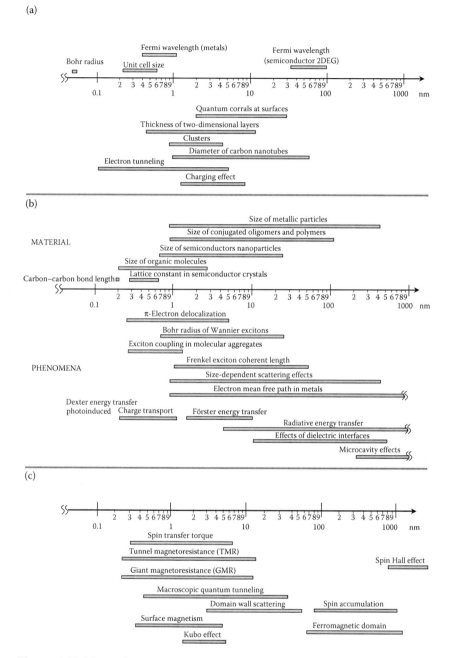

Figure 1.23 Material properties with respect to characteristic size of nano-structures: (a) eletronic properties; (b) optical properties; (c) magnetic properties.

In crystalline materials, a regular arrangement of atoms is represented by repeating unit cells. In conventional materials such as normal metals, the unit cell includes several atoms and has a size of several tens of nanometers. Meanwhile, the Fermi wavelength of the valence electron is also of the order of several tens of nanometers in conventional materials. This Fermi wavelength is much smaller in comparison with the usual size of materials (ca. millimeter or centimeter scale). In this case, the effect of the electron confinement on the material's electronic states is negligible. However, for materials with sizes on the nanometer scale, the electron confinement greatly modifies the intrinsic electronic states of the materials. The nanometer-sized materials start to reveal their quantum nature. The diameter of carbon nanotubes is depicted as a typical size of nanoscale materials in the figure. In Chapter 2, we consider several examples of quantum effects induced by the 1D, 2D, and 3D electron confinements of the electrons in such nanoscale materials including the quantum corrals at materials surfaces, 2D electron systems in ultrathin films, nanoscale clusters, and so on. In addition to electron confinement, electron tunneling and charging effects become pronounced in materials on the nanometer scale. These topics are also introduced in the next chapter. The aforementioned description suggests that the materials of larger sizes will also show the quantum nature if their Fermi wavelength is long enough. Indeed, this situation can be realized, for example, in the 2D electron systems at semiconductor interfaces. These systems are known as *mesoscopic*. However, mesoscopic systems must be cooled to very low temperatures in order to observe size-driven quantum effects. This is because the quantum states of larger systems are destroyed by thermal fluctuation at ambient temperatures. In Chapter 2, we do not treat mesoscopic phenomena, but briefly discuss the effect of size on the robustness of the quantum nature of materials. This discussion will shed light on the advantages of nanoscale materials.

As described in Chapter 3, the optical properties and light–matter interactions change profoundly as the size of materials decreases to nanometer scales. The typical length scales in materials and physical phenomenon related to optical properties are shown in Figure 1.23b. Some of the most dramatic changes in the optical properties of materials that are perceived by humans as changes of materials color occur on the scales 0.1–10 nm. When considering size effects in materials, we may begin by realizing the size of the *building blocks* of such materials. In inorganic crystalline materials, this could be a unit cell characterized by the dimensions of the lattice constant. For most common inorganic semiconductors, for example, the lattice constant is in the order of 0.3–0.7 nm. In organic materials, the basic building block could be a carbon–carbon bond with the length of 0.12–0.16 nm. Optical absorption and emission properties of organic materials are determined by the delocalization length of π-electron, which is given by the size of the molecule in the order of 0.2 to several nanometers; in conjugated polymers, the delocalization length can be still larger. The properties can be further changed by correlated absorption and emission of many molecules in molecular aggregates or molecular crystals—the related effects are described by the Frenkel exciton and its coherent length, which can range from a few to tens of nanometers. In inorganic semiconductors, the absorption and emission properties change most when the size of the crystal is comparable or smaller than the size (Bohr radius) of the electron–hole pair, the Wannier exciton. Typically, the exciton Bohr radius is between 0.7 and 25 nm for most semiconductors. In metallic materials, changes of optical properties due to adsorption and scattering are observed in a wide range of sizes, from subnanometers to several hundreds of nanometers. Another relevant property of small metallic particles, the electron mean free path, ranges from a few nanometers to macroscopic scales, depending on the metal purity and temperature. Size-dependent interaction between material nano-objects can be either short-range (subnanometers) if they are due to overlap of atomic or molecular orbitals (e.g., energy and electron transfer) or long-range (several to tens of nanometers) if they are due to dipole–dipole or other electronic interactions, such as Förster energy transfer or effects of dielectric interfaces. The relevant scale for changes of light–matter interaction in confined geometries (micro- and nano-cavities) is the half-wavelength of light, which, depending on the color, is between 200 and 360 nm.

The relevant length scales of notable magnetic and magnetotransport phenomena notable in nanostructures, which are discussed in Chapter 4, are depicted in Figure 1.23c. In principle, the Kubo effect, macroscopic quantum tunneling, and surface magnetism are associated with the modified electronic states due to the confinement of electrons in nanoclusters and spin-dependent DOS at the very top of the surface so that their characteristic length scale ranges between 0.1 and 10 nm. In terms of magnetotransport properties, the mechanism of giant magnetoresistance (MR) effect is closely associated with spin-dependent relaxation time of electrons at the ferromagnetic/nonmagnetic (FM/NM) metal interface, while that of tunneling MR is a spin-dependent tunneling process of electrons across the FM metal/insulator/FM metal junction (0.1 to several nanometers). Another important aspect is the diffusion process of electron spins in NM metals, providing spin accumulation and the spin Hall effect. These diffusion processes are characterized by spin diffusion length in the NM metals (submicrometers to a few micrometers). Also, FM domains (micrometers) and FM domain walls (~10 nm) give rise to marked features in magnetic and magnetotransport phenomena.

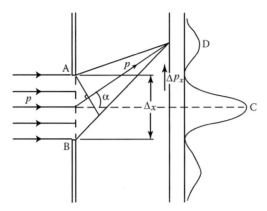

Figure 1.24 Electron diffraction at slit AB and the intensity distribution of the wall.

PROBLEMS

1.1 Figure 1.24 shows the diffraction of electrons at slit AB and the intensity distribution of the wall. Using the information included in this figure, derive the relationship $\Delta p_{x0} \Delta x = h$, where h is Planck's constant, Δp_x is the uncertainty of momentum, and Δx is the uncertainty of position.

1.2 Show that the eigenvalue of a Hermitian operator is a real number.

1.3 Describe Bragg's law of diffraction using real and reciprocal unit vectors.

1.4 Find the relationship for the variation of the rate of wave number, dk/dt, for an electron moving in a crystal under a force F.

1.5 Describe the similarities and differences between the Hartree–Fock approximation and the X_α method.

References

1. R. Feynman, R. Leighton, and M. Sand, *The Feynman Lectures on Physics, Vol. 3*, Addison-Wesley, Massachusetts, 1964.
2. P. Dirac, *The Principles of Quantum Mechanics*, 4th Ed., Oxford University Press, London, 1958.
3. I. Bunget and M. Popescu, *Physics of Solid Dielectrics*, Elsevier, Amsterdam, 1984.
4. C. Kittel, *Introduction to Solid State Physics*, 7th Ed., Wiley, Hoboken, 1995.
5. J.C. Slater, Statistical exchange—correlation in the self-consistent field, *Adv. Quantum Chem.* 6, 1972, 1.
6. W. Kohn and L. Sham, Self-consistent equations including exchange and correlation effects, *Phys. Rev.* 140, 1965, A1133.

7. R.M. Dreizler and E.K.U. Gross, *Density Functional Theory—An Approach to the Quantum Many-Body Problem*, Springer, Berlin, 1990.
8. R.G. Parr and W. Yang, *Density-Functional Theory of Atoms and Molecules*, Oxford Science Publications, New York, 1989.
9. P. Fulde, *Electron Correlations in Molecules and Solids*, 2nd Ed., Springer, Berlin, 1993.

chapter two

Electronic states and electrical properties of nanoscale materials

2.1 Outline

Reduction in the size of a specimen of a material in a specific direction, for example, the z-direction, restricts the freedom of motion of electrons along the z-axis, although the electrons can still move freely in the x–y-plane. Now, if the size of the specimen along the z-axis is of the same order of magnitude as the wavelength of the electron, namely the nanoscale, then the specimen assumes a truly nanoscale sheet shape and has a 2D electronic structure. Furthermore, the specimen becomes a nanowire and its electronic states have 1D character with reduction in the size along the x- and y-axes. Reducing the size along all the three axes yields a specimen known as a quantum dot, in which the electronic states have 0D character. Reducing the size of materials to the nanometer scale lowers their dimensionality and modifies the electronic structure (Figure 2.1).

Electrons traveling along the direction that has been reduced in size reach the boundaries, that is, the surfaces of the specimen, and are confined in nanospace. This confinement causes the quantization of the electron wavelength and the energy spectrum. In addition, the cyclic boundary condition is broken at the edges (surfaces), and the contribution of the edge-localized electronic states (surface states) dominates the total electronic structure. The electron-charging energy also increases dramatically due to the size-dependent increase in the electrostatic capacitance of nanoscale materials. Furthermore, reduction in the size leads to the manifestation of electron tunneling phenomena.

The basic concepts of these size effects, which dominate the physical properties of nanoscale materials, are described in the first part of this section. Next, a concise picture of two factors that hinder the manifestation of the quantum nature of nanomaterials is introduced. In general, the smaller the size, the more clearly quantum size effects can be observed in nanomaterials. However, the effects are frequently hidden by thermal

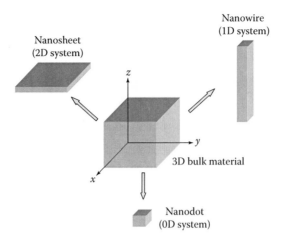

Figure 2.1 Formation of nanostructures by reducing the size in one, two, and three dimensions.

broadening and decoherence. The size necessary to overwhelm these destabilizing factors is also discussed.

Finally, the structural stability of nanomaterials is discussed from the viewpoint of electronic energy. Nanomaterials fall into local minima in electronic energy at specific (magic) shapes and sizes in the course of reducing the size of materials. This is technologically useful in controlling the size and shape of nanomaterials by the so-called self-assembly. The basic ideas of self-assembly and some pertinent examples are described.

In this chapter, we assume *a priori* that the material is crystalline and that its electronic structure is described well by conventional band theory for topics where energy dispersion is important. However, in some cases, materials are amorphous or polycrystalline, that is, they do not have a crystalline structure. Furthermore, band theory fails to describe the electronic structure even in crystalline materials referred to as strongly correlated systems.

2.2   Low dimensionality and energy spectrum

2.2.1   Space for electrons in materials

In crystalline materials, unit cells (typically 0.5 nm in size) pile up in a regular fashion along all the x-, y-, and z-directions. For macroscopic specimens of ~5 mm in size, there are ~10^7 unit cells located along each direction, and the Fermi wavelength of valence electrons is of the same order as the unit cells, for example, in conventional metallic materials.

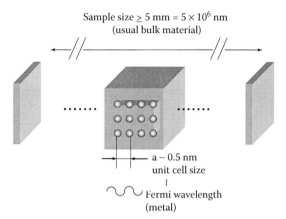

Figure 2.2 Scale of a sample unit cell, and the Fermi wavelength.

Thus, the inner space is too large for electrons to feel the effects of boundaries in usual macroscopic materials. The electrons are moving in an infinitely wide space and do not reach the boundaries of materials (Figure 2.2).

As we learned in Section 1.2, the infinite space for electrons is mathematically expressed by the periodic boundary conditions for materials on the scale $L \gg a$ (a is the lattice constant) in the standard theory of solid-state electronic structure. This restricts the possible k-vectors to be $k = (2\pi/L)l$, where l is an integer. Thus, the Brillouin zone $-\pi/a < k < +\pi/a$ includes an incredibly large number ($L/a \sim 10^7$) of k inside it. Consequently, the possible energy levels $\varepsilon(k) = (\hbar^2/2m)k^2$ are dense with very small energy spacing for the electronic bands in the energy spectrum. This is the reason why macroscopic-sized materials exhibit electronic band structure.

2.2.2 Electron DOS of 3D materials with macroscopic dimensions

Here we consider a crystalline sample, the size of which is larger in all the x-, y-, and z-directions than the electron wavelength. We assume that electrons move freely along all the three dimensions. In this case, assuming a mass m, wavevector $k = (k_x, k_y, k_z)$, and Planck's constant $\hbar$, the electron energy $E_{3D}(k)$ is given by

$$E_{3D}(k) = \frac{\hbar^2}{2m}(k_x^2 + k_y^2 + k_z^2) = \frac{\hbar^2}{2m}k^2. \tag{2.1}$$

Under periodic boundary conditions for a specimen of size L, k takes the values $[(2p/L)l_x, (2p/L)l_y, (2p/L)l_z]$ in reciprocal space, where l_x, l_y, and l_z are integers.

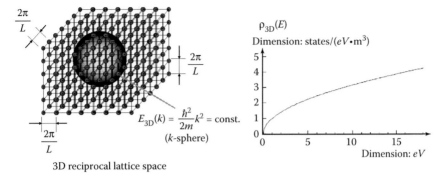

3D reciprocal lattice space

Figure 2.3 **(See color insert following page 148.)** The Fermi sphere in the 3D reciprocal lattice (left) and the electron DOS (per energy) of the 3D crystalline material (right).

Electrons with energies $E \leq E_{3D}(k)$ occupy the k-states inside a sphere of radius k in reciprocal space, where $E_{3D}(k) = (\hbar^2/2m)k^2$. Since each k-state can be occupied doubly by electrons with up and down spin, the number of electrons $N_{3D}(E)$ in a sphere of radius k is given by

$$N_{3D}(E) = \frac{2 \times \left(\int_0^k 4\pi k^2 \, dk \right)}{(2\pi/L)^3} = \frac{L^3}{3\pi^2} k^3 = \frac{L^3}{3\pi^2 \hbar^3} (2mE_{3D})^{3/2}. \qquad (2.2)$$

Now, the electron density per volume $n_{3D}(E)$ is given by $n_{3D}(E) = N_{3D}(E)/L^3$. Thus, as we saw in Section 1.2, the electron DOS (per energy) is determined to be

$$\rho_{3D}(E) = \frac{\partial n_{3D}(E)}{\partial E} = \frac{(2m)^{3/2}}{2\pi^2 \hbar^3} \sqrt{E}. \qquad (2.3)$$

This equation shows that for macroscopic 3D crystalline materials, the electron DOS is proportional to $\sqrt{E}$ (Figure 2.3).

AN ASIDE: HOW DO WE INDEX ELECTRONS IN SOLIDS?

In classical physics, electrons are treated as being tiny particles. Thus, in solids, electrons are expected to be distributed as the points shown in the left panel of Figure 2.4. Here, we can index each electron by its position $\mathbf{r}$.

On the other hand, in quantum mechanics, the behavior of electrons is expressed in terms of a wave function ψ. Specifically, the wave function has the form of a plane wave $\psi(\mathbf{r}) \propto e^{i\mathbf{k}\cdot\mathbf{r}}$ for electrons moving freely in solids. The laws of quantum mechanics dictate that

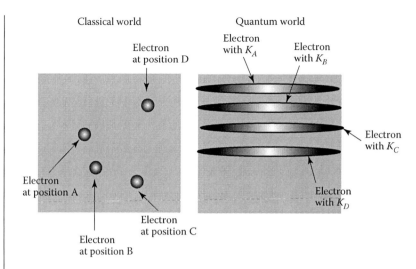

Figure 2.4 Indexing the electrons in the classical (left) and in the quantum physics (right).

the probability of finding an electron at a position **r** is given by $|\psi(\mathbf{r})|^2$. However, $|\psi(\mathbf{r})|^2$ becomes a constant for the wave function $\psi(\mathbf{r}) \propto e^{i\mathbf{k}\cdot\mathbf{r}}$. This implies that electrons are completely delocalized in solids and thus cannot be indexed by their position.

So how do we index electrons in solids? The answer is by using the wavevector **k**. Under periodic boundary conditions, the wavevector **k** is strictly defined by a combination of integers (l_x, l_y, l_z) as $k = [(2\pi/L)l_x, (2\pi/L)l_y, (2\pi/L)l_z]$. If N unit cells, each with a dimension a, are included in a specimen of size L (i.e., $L = Na$), then the Brillouin zone $(-\pi/a < k_x < \pi/a, -\pi/a < k_y < \pi/a, -\pi/a < k_z < \pi/a)$ contains $2N^3$ states inside it. Here, the factor of 2 comes from the spin degeneracy at each state. On the other hand, the specimen contains N^3 electrons, if, for example, each unit cell ejects one valence electron. Thus, all the N^3 electrons can be indexed by the **k**-vector in the Brillouin zone.

The wavevector **k** is related to the momentum **p** by $\mathbf{p} = \hbar\mathbf{k}$. Between the fluctuation of $\mathbf{p}(\Delta\mathbf{p})$ and $\mathbf{r}(\Delta\mathbf{r})$, the uncertainty principle $\Delta\mathbf{p} \cdot \Delta\mathbf{r} \geq \hbar/2$ is valid, and in the present case, $\Delta\mathbf{p} = 0$ and $\Delta\mathbf{r} = \infty$. Thus, the electrons are completely delocalized (i.e., $\Delta\mathbf{r} = \infty$), but are indexed by **k** (i.e., $\Delta\mathbf{p} = 0$).

2.2.3   Electron DOS in 2D materials (nanosheets)

In contrast to macroscopic-sized materials, the size of nanomaterials L_{nano} is the same or slightly larger than the electron wavelength, at least along one of the three orthogonal axes. This means that on the nanometer scale, electrons reach the boundaries and feel confined in a narrow space. Along a given nanometer-scale direction, the electron wavelength λ is quantized to satisfy the Bohr–Sommerfeld condition $L_{nano} = (\lambda/2) \times n$. Here n is an integer called the quantum number. By a careful examination of the relationship $k = 2\pi/\lambda = (\pi/L_{nano})n$ for $L_{nano} \ll L$ (where L is the size of a macroscopic material), we realize that the k-vector is much more dispersed in the nanometer-scale direction than along the macroscopic scale direction $[k = (2\pi/L)l]$. As a result, the energy levels $E(n) = (\hbar^2/2m)k^2$ become discrete in order to manifest their quantum nature along the nanometer-scale direction. In the following section, we consider how the quantization of nanomaterials modifies their electron DOS.

We first consider the simplest case of a *nanosheet*, that is, an infinitely wide thin film (in the x–y-plane) with a nanometer thickness d (along the z-direction) as shown in Figure 2.5. Electrons move freely along the x–y-plane in the thin film, whereas the motion is restricted and the kinetic energy is quantized into discrete quantum levels $E_z(n_z)$ with quantum number n_z, along the z-direction. In this case, the total energy of electrons is given by

$$E(k_{2D}) = \frac{\hbar^2}{2m}(k_x^2 + k_y^2) + E_z(n_z). \tag{2.4}$$

Here, $k_{2D} = (k_x, k_y)$ is the wavevector of electrons moving freely in the x–y-plane. $E_z(n_z)$ is the quantized discrete energy level with a quantum number n_z for motion along the z-axis. In the k_x–k_y reciprocal plane, electrons with energies below $E(k_{2D})$ occupy the states $k_{2D} = (k_x, k_y)$ inside a circle of radius

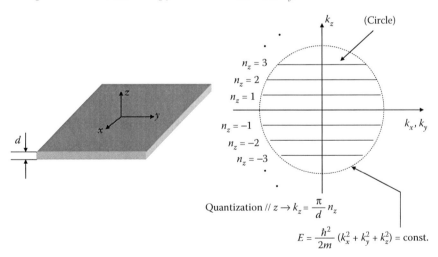

$$\text{Quantization} \,/\!/\, z \to k_z = \frac{\pi}{d}\, n_z$$

$$E = \frac{\hbar^2}{2m}(k_x^2 + k_y^2 + k_z^2) = \text{const.}$$

Figure 2.5 Counting the k-states in the 2D nanosheet.

$k \leq \sqrt{2m}/\hbar\sqrt{E(k_{2D}) - E_z(n_z)}$ for a given quantized state $E_z(n_z)$. Since the periodic boundary conditions with L are still valid for free-electron motion along the x- and y-directions, then the k_{2D}-state takes values $[(2\pi/L)l_x, (2\pi/L)l_y]$ in the k_x–k_y-plane (l_x and l_y are integers). Therefore, the number of electrons (the number of states in the circle in the k_x–k_y reciprocal plane) is counted as

$$N_{2D}(E) = \frac{2 \times \left(\int_0^k 2\pi k \, dk \right)}{\left(2\pi/L\right)^2} = \frac{L^2}{2\pi} k^2 = \frac{L^2}{2\pi} \frac{2m}{\hbar^2} \{E(k_{2D}) - E_z(n_z)\}. \tag{2.5}$$

The factor 2 in the above equation comes from the spin degeneracy at each k_{2D}-state. The electron *sheet* density is $n_{2D}(E) = N_{2D}(E)/L^2$. Thus, the electron *sheet* DOS (per unit energy) is given by

$$\rho_{2D}(E) = \frac{\partial n_{2D}(E)}{\partial E} = \frac{1}{2\pi} \frac{2m}{\hbar^2}. \tag{2.6}$$

In 2D nanosystems, the conventional $\sqrt{E}$ dependence of $\rho_{3D}(E)$ for macroscopic 3D materials is independent of E.

In the above calculation, we considered the DOS of a 2D system belonging to a quantized electronic state. However, electrons can take a range of quantized states along the z-direction. Repeating the calculation described above, we find that each quantum level $E_z(n_z)$ (along the z-direction) supports the same sheet electron DOS $\rho_{2D} = (1/2\pi)(2m/\hbar^2)$. Thus, the total electron *sheet* DOS is given by the product of $\rho_{2D} = (1/2\pi)$ $(2m/\hbar^2)$ and the number of quantum levels available along the z-direction. Increasing the electron energy, E, leads to a stepwise increase in the number of quantum levels $E_z(n_z) \leq E$. Thus, the total electron *sheet* DOS changes like an upward staircase as a function of the electron energy E, as illustrated in Figure 2.6. In a 2D nanosystem, the continuous $\sqrt{E}$ dependence of the DOS for macroscopic 3D materials is split into *subbands* associated with each of the quantum well states (QWSs) along the z-direction.

The existence of a subband structure in 2D nanosystems can be experimentally verified. An example is shown in Figure 2.7, where a 16-monolayer (ML)-thick film of Ag was deposited on a Si(111) substrate (1 ML of Ag is equal to 0.236 nm). By depositing at a low temperature and subsequently slowly annealing to room temperature, the film possesses an atomically flat surface morphology as shown in the upper left scanning tunneling microscope (STM) image in the figure [1]. The resulting nanometer-thick, atomically flat Ag film can be regarded as a 2D nanosheet system. Bulk Ag valence electrons exhibit free-electron-like energy dispersion. However, in the ultrathin film, the electrons are confined and generate quantum well

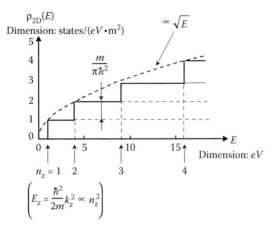

Figure 2.6 Electron DOS of the 2D nanosheet.

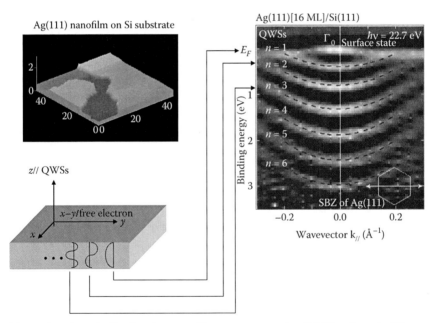

Figure 2.7 Atomically flat Ag nanofilms grown on the Si(111) substrate. Upper left: STM image of the grown Ag nanofilm. Lower left: a schematic of the electronic states in the Ag nanofilm. Right: the subband structure observed by an ARUPS measurement. (Reprinted from I. Matsuda, T. Ohta, and H.W. Yeom, *Phys. Rev. B* 65, 2002, 085327. With permission.)

states along the thickness (growth) direction, as illustrated in the lower left panel in Figure 2.7, although they still maintain a free-electron-like freedom of motion in the plane direction. For electrons with energies below the Fermi level, the energy dispersion is successfully observed using angle-resolved ultraviolet (UV) photoelectron spectroscopy (ARUPS) as shown in the right panel in Figure 2.7 [2]. The vertical and horizontal axes in the figure indicate electron energies below the Fermi level and the in-plane wavevector $|k_{2D}|$, respectively. The ARUPS spectrum here shows several free-electron-like subband dispersions. Each of the dispersion curves originates from energies equal to QWSs (of the quantum number $n = 1$, 2, 3,...) at the Γ point (namely, at $|k_{2D}| = 0$). For increasing $|k_{2D}|$, the energy increases as $\hbar^2/2m\,|k_{2D}|^2$ in each branch, as expected for free-electron-like in-plane motion in 2D nanomaterial systems. This is direct evidence of subbands in 2D nanosystems, namely free-electron motion along the x–y-plane accompanied by the quantum well states along the z-direction in atomically flat, nanometer-thick Ag films.

AN ASIDE: ARUPS

Characterization of 2D electron band dispersions is important in modern materials science and engineering. ARUPS is one of the most powerful methods for such purposes. Figure 2.8 (left panel) is an illustration of the experimental setup of ARUPS, where the sample surface is irradiated by monochromatic UV light with a photon energy $\hbar\nu$. The valence electrons make transitions in excited states on absorption of the UV light. If the energy of the excited electrons exceeds the vacuum level E_{vac}, then these electrons are ejected from the sample and can be detected using electron detectors. Typically, the electron detector is set to monitor electrons possessing a kinetic energy E_{kin}, ejected in the direction θ from the normal to the surface. By the detected energy E_{kin}, we can estimate the original energy position E of the emitted electrons with respect to the Fermi level (E_f) according to the equation $E = \hbar\nu - \Phi - E_{kin}$, which is based on the energy relationship shown in the right panel in Figure 2.8. Here, Φ is the work function, that is, the energy difference between the vacuum and the Fermi levels of the sample. The intensity of the emitted electrons is proportional to the original DOS in the valence band. Furthermore, the wavevector parallel to the surface $|k_{2D}| = \sqrt{((2mE_{kin})/\hbar^2)}\sin\theta$ is conserved in the optical transitions. Thus, by sweeping the detecting energy and angle of the electron detector, ARUPS enables us to resolve the 2D energy dispersion, that is, the relationship between $E(k_{2D})$ and $|k_{2D}|$ of the valence electrons.

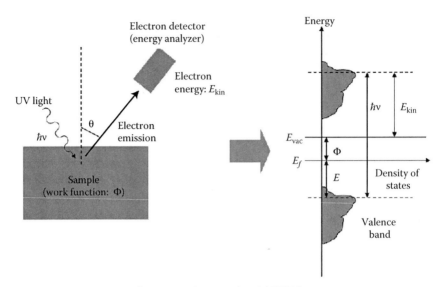

Figure 2.8 Experimental setup and principle of ARUPS.

2.2.4   *Electron DOS in 1D materials (nanowires)*

Reduction in the size in the x- and y-directions results in materials known as nanowires. In the x–y-plane, electrons are confined and quantized into discrete energy states $E_{x,y}(l, m)$, where l and m are integer quantum numbers used to express the quantized states. If the cross-section of a wire is rectangular in shape, then the quantum numbers l and m are related to the two sizes of the cross-section (a in the x-direction and b in the y-direction) as

$$a = l\frac{\lambda_x}{2}, \quad \text{and} \quad b = m\frac{\lambda_y}{2}, \tag{2.7}$$

where λ_x and λ_y are the electron wavelength along the x- and y-directions (Figure 2.9).

In the case of cylindrical nanowires, l and m denote the principal radial and angular momentum quantum numbers, respectively. Generally, the quantized states in the cross-section of nanowires are described by a combination of two quantum numbers (l, m).

In nanowires, the total energy of electrons $E(k_z)$ is given by the sum of their kinetic energy of free motion along the z-axis and the quantized energy in the x–y-plane as

$$E(k_z) = \frac{\hbar^2}{2m}k_z^2 + E_{x,y}(l, m), \tag{2.8}$$

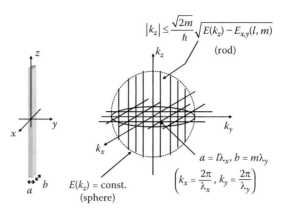

Figure 2.9 Counting the k-states in the 1D nanomaterials.

where k_z is the wavevector of electrons along the z-(axial) direction. Here, we assume the length of the nanowire to be infinite compared with the size of its cross-section. This assumption is mathematically treated by adapting a cyclic boundary condition along the z-direction, which results in $k_z = (2\pi/L)l_z$, where L is the length of the wire and l_z is integer $\pm 1, \pm 2, \pm 3, \ldots$ Thus, the number of electrons of energy below $E(k_z)$ with a quantum state $E_{x,y}(l, m)$ is determined as

$$N_{1D}(E) = \frac{2 \times \int_{-k_z}^{k_z} dk_z}{2\pi/L} = \frac{2L}{\pi} k_z = \frac{2L}{\pi} \frac{\sqrt{2m}}{\hbar} \sqrt{E(k_z) - E_{x,y}(l, m)}. \quad (2.9)$$

Here, note that the integral for k_z is carried out over the range $(-k_z, k_z)$, since electrons can move freely in both the $+z$- and $-z$-directions in the nanowire. The electron density *per unit length* along the z-axis is $n_{1D}(E) = N_{1D}(E)/L$. Therefore, the electron DOS (per unit energy) is

$$\rho_{1D}(E) = \frac{\partial n_{1D}(E)}{\partial E} = \frac{\sqrt{2m}}{\pi\hbar} \frac{1}{\sqrt{E - E_{x,y}(l, m)}}. \quad (2.10)$$

The electron DOS shows a singularity as $1/\sqrt{E}$ in the 1D nanowire (Figure 2.10).

The above result is associated with a QWS in the x–y cross-section. However, in nanowires, many QWSs are available in their cross-section. Therefore, the total electron DOS is given by summation of $n_{1D}(E)$ for

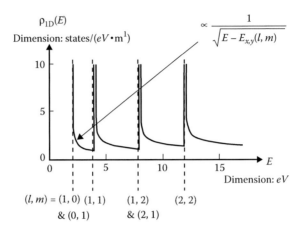

Figure 2.10 Electron DOS in 1D materials (nanowires).

possible quantum states in the x–y-plane. Finally, the electron DOS of 1D nanowire systems is given by

$$\rho_{1D}(E) = \sum_{l,m} \frac{\sqrt{2m}}{\pi\hbar} \frac{1}{\sqrt{E - E_{x,y}(l, m)}},\tag{2.11}$$

where the summation is taken for (l, m) of $E_{x,y}(l, m) < E$.

2.2.5   *Quantized conductance in 1D nanowire systems*

The electron states associated with quantized states (l, m) are called channels in a 1D system such as nanowires. To illustrate the role of channels, next we consider the electron transport in 1D systems (Figures 2.11 and 2.12).

First, we neglect quantum states in the x–y-plane (i.e., the size of the x–y-plane), and consider electron transport in a purely 1D system such as a needle along the z-direction as shown in Figure 2.11, where both sides of the wire are connected to metal electrodes, and a bias voltage is applied between them in order to pass an electron current through the 1D nano-wire system.

To calculate the magnitude of the current, we denote the chemical potentials of the electrodes as μ_R and μ_L. Here the suffixes L and R mean properties belonging to the left and right electrodes. As illustrated in this figure, application of a positive bias voltage V to the right electrode lowers μ_R by eV with respect to the chemical potential, μ_L of left electrode. This enables the transport of electrons from the occupied states in the left-side electrode to empty states in the right-side one. For simplicity, we assume

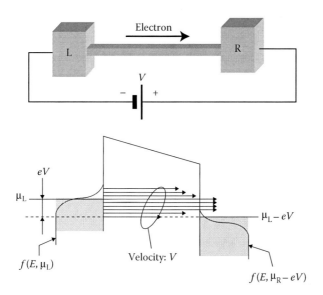

Figure 2.11 Electron transport in a 1D nanowire. Upper: experimental setup. Lower: energy band diagram.

that electrons move with a group velocity $v(k)$ and a transmission probability $T(k)$ in the nanowire, where k is the wavevector of the electron along the z-axis. Now, the number of electrons η, which pass the cross-section of the nanowire in a time duration τ, is given by the length of the segment $v \cdot τ$, for electrons that pass the cross-section in the time duration τ multiplied by the density of electrons per unit length in the 1D nanowire. As we discussed in the previous section, the density of electrons in the nanowire (per unit length) is given by $1/L \cdot 1/(2π/L) \times 2$. Again, the factor

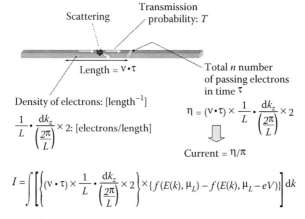

Figure 2.12 Calculation of the conductance in the 1D nanowire.

2 here comes from the spin degeneracy in each k-state. Finally, η is given by $\eta = (v \cdot \tau) \times 1/L \cdot 1/(2\pi/L) \times 2$ (Figure 2.12).

In the above calculation, we assumed k to be a constant along the z-direction. This is equivalent to fixing the energy of the electrons moving in the wire. However, as illustrated in Figure 2.11, electrons with energies between μ_L and $\mu_R - eV$ can contribute to the electron transport. And, generally, the transmission probability is also a function of k. Therefore, based on these factors, we obtain an expression for the total current $I = \eta/\tau$ as

$$I = 2e \int_0^\infty \{f(E(k), \mu) - f(E(k), \mu - eV)\} v(k) T(k) \frac{dk}{2\pi}. \tag{2.12}$$

To change the integral in Equation 2.12 from E to k, we substitute the following relationship into Equation 2.12:

$$dk = \frac{dk}{dE} dE = \frac{1}{\hbar v} dE, \tag{2.13}$$

and obtain

$$I = \frac{2e}{2\pi\hbar} \int_0^\infty \{f(E(k), \mu) - f(E(k), \mu - eV)\} T(E) dE$$

$$= \frac{2e}{h} \int_0^\infty \{f(E, \mu) - f(E, \mu - eV)\} T(E) dE. \tag{2.14}$$

Here, $f(E, \mu)$ and $f(E, \mu - eV)$ are the Fermi distribution functions in the left and right electrodes. Furthermore, in cases where the bias voltage V is sufficiently small, the following equation holds for the Fermi distribution function:

$$f(E(k), \mu) - f(E(k), \mu - eV) \cong eV \frac{\partial f(E, \mu)}{\partial \mu} = -eV \frac{\partial f(E, \mu)}{\partial E}. \tag{2.15}$$

Substituting Equation 2.15 into Equation 2.14, we finally obtain a simple expression for the conductance $G = I/V$ in the nanowire as

$$G = \frac{2e^2}{h} \int_0^\infty \left\{ -\frac{\partial f(E, \mu)}{\partial E} \right\} T(E) dE \equiv \frac{2e^2}{h} T_\mu. \tag{2.16}$$

In deducing Equation 2.16, we used the relationship $-\partial f(E, \mu)/\partial E = \delta(E - \mu)$, because the Fermi distribution function $f(E, \mu)$ changes in a stepwise fashion at $E = \mu$ at $T = 0$ K. Equation 2.16 implies that in 1D systems, the conductance of electrons is quantized in units of $2e^2/h$.

In this calculation described, we ignored the thickness of the wire, in spite of the fact that real 1D nanowires do have finite thicknesses. Therefore, the electron states described by k in the z-direction are actually associated with possible quantum states (l, m) in the x–y-plane. Since the calculation of the electron conductance was independent of the choice of quantum states in the x–y-plane, we readily obtain the result that each of the quantum states (l, m) has equal unit conductance $(2e^2/h)T_\mu$ along the wire axis. Thus, the conductance of nanowires with finite thicknesses is given by the product of the number of available quantum states (l, m) below the Fermi level and their quantized conductance $(2e^2/h)T_\mu$. In this definition, each quantum state (l, m) is regarded as a 1D conductance channel that supports the unit quantum conductance $(2e^2/h)T_\mu$ along the axis of the wire.

The quantization of conductance and the contribution of channels in 1D nanowires are actually observed in experimental measurements of nanowires. An example is shown in Figure 2.13 [3]. In this experiment, a gold nanowire is constructed by forming a nanocontact and subsequent retraction of an atomically sharpened tungsten tip at the gold substrate surface inside an STM system. This procedure results in the formation of a gold 1D nanowire in the gap between the tip and the surface of the gold, as shown in the right side of the figure. Retracting the tip elongates the gold wire and causes a reduction in its radius, just as you would find by stretching out a chewing gum. The conductance of the gold nanowire is measured as a function of the tip-to-surface distance during this retraction. As shown in the figure, the conductance decreases with the tip–surface distance. However, the change of the conductance is not continuous, but stepwise. The plateau in the tip–surface distance versus the conductance curve corresponded with multiples of quantum conductance units as we expect from the calculations described above. The multiplication factor corresponds to the number of available channels. This experimental result is understood as follows: in the course of the tip retraction, the cross-section of the wire is reduced and the energy levels of the QWSs in the x–y cross-section increase and disperse. This causes the decrease in the number of available channels below the Fermi level, and the current decreases stepwise at each missing channel during the tip retraction.

2.2.6   Electron DOS in 0D materials (nanodots)

Reduction in the size of materials in all three dimensions results in the formation of 0D systems called *quantum dots* or *nanodots*. In quantum dots,

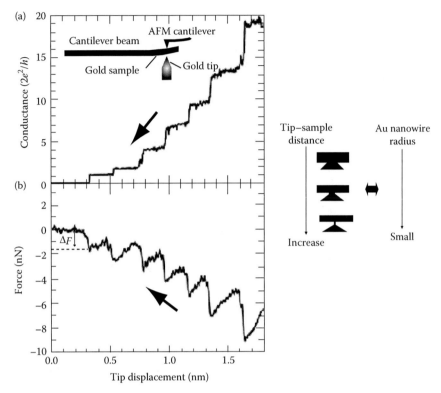

Figure 2.13 Experimental evidence of the quantization of conductance in the 1D nanowire. (a) The conductance decreased stepwise at each missing of the available channels in the tip retraction. (b) The stepwise change of the force between the tip and the sample. (Reprinted from G. Rubio, N. Agrait, and S. Vieira, *Phys. Rev. Lett.* 76, 1996, 2302. With permission.)

the electron is fully quantized in the x-, y-, and z-directions with quantum numbers l, m, and n, and the electron DOS is given by

$$\rho_{0D}(E) = 2 \sum_{l,m,n} \delta(E - E_{l,m,n}), \qquad (2.17)$$

where $E_{l,m,n}$ is the quantized energy level along the x-, y-, and z-directions (Figure 2.14). The δ-function appears for the following reasons. In nanodots, electrons are allowed to exist only at energies equal to $E_{l,m,n}$. This behavior is represented by the δ-function $\delta(E - E_{l,m,n})$. The relationship $\int \delta(E - E_{l,m,n}) \, dE = 1$ guarantees that each $\delta(E - E_{l,m,n})$ accommodates one electron state at $E = E_{l,m,n}$. The factor of 2 in Equation 2.17 is due to the spin

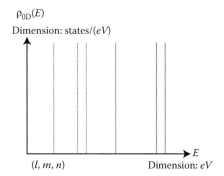

Figure 2.14 Electron DOS in 0D materials (nanodots).

degeneracy, and the summation is taken over all the quantized states that satisfy the relation $E > E_{l,m,n}$.

From the perspective that the energy levels are fully discrete, the nanodot can be envisaged as artificial atoms. Thus, electrons in nanodots can absorb photon energy by the photo-excited electron transition from the highest occupied state to the lowest unoccupied state, thus mimicking the optical absorption behavior of atoms. Since the wave functions of excited states fully overlap with those of ground states in the narrow confined space in nanodots, excited electrons recombine with the holes remaining at ground states, and the emission of light is extremely efficient. Here, the most prominent characteristic of nanodots is that the energy gap between the electron states varies with dot size. This means that the wavelength of the emitted light, that is, the color of photoluminescence, can be tuned artificially by changing the size of the nanodot. A more detailed discussion on the optical properties of nanodots will be given in Section 3.1.

2.3   Quantization

In the previous subsection, we concluded that quantum confinement modifies the electron DOS and manifested the unique physical characteristics of materials with nanometer-scale dimensions. On the other hand, recent progress in nanoscience and technology enables methods for fabricating artificial nanostructures. Thus we have the technology for freely tailoring the electronic properties of materials by controlling the structure of nanomaterials. The key point here is to design the shape and size of the nanostructure and to control the energy levels and wave functions of the QWSs in order to produce nanomaterials exhibiting the desired functions.

In this subsection, we consider the artificial design of electron states of materials, and first examine the effect of the shape of nanostructures for 2D square and circular quantum wells with the same size. Then, we see how a realistic finite confinement potential height modifies the energy levels of QWSs calculated assuming a simple, but unrealistic infinite confinement potential height. We also describe a way of including band dispersion in the calculation of the confined QWSs. Numerically calculated figures and tables are presented as a guide for a general understanding of the scale of these effects in nanoworld.

2.3.1   2D square wells

First, we consider the electron states in a 2D square well with an infinite potential wall. But before discussing the 2D well, let us briefly review a 1D quantum well system bounded by infinite potential barriers. In a 1D quantum well with infinite potential barriers and a well width of d (i.e., $-d/2 < x < d/2$), the wave functions and energies are quantized and are given by

$$\psi_n(x) = \sqrt{\frac{2}{d}} \cos\left(\frac{n\pi x}{d}\right) \quad \text{for odd values of } n, \qquad (2.18a)$$

$$\sqrt{\frac{2}{d}} \sin\left(\frac{n\pi x}{d}\right) \quad \text{for even values of } n, \qquad (2.18b)$$

$$E_n = \frac{\hbar^2}{2m}\left(\frac{n\pi}{d}\right)^2. \qquad (2.19)$$

In a 2D square well, the confined electrons are quantized in both the x- and y-directions. In a 2D square well, since the quantization occurs along the x- and y-directions independently, the wave functions are given by the product of the quantized states in the x- and y-directions as

$$\psi_{l,n}(x, y) = \phi_l(x)\phi_n(y). \qquad (2.20a)$$

Components of $\phi_l(x)$ and $\phi_n(y)$ are determined in the same way as 1D quantum well systems and are given by

$$\phi_l(x) = \sqrt{\frac{2}{d}} \cos\left(\frac{l\pi x}{d}\right) \quad l: \text{ odd,}$$

$$\phi_l(x) = \sqrt{\frac{2}{d}} \sin\left(\frac{l\pi x}{d}\right) \quad l: \text{ even,} \qquad (2.20b)$$

and

$$\phi_n(y) = \sqrt{\frac{2}{d}} \cos\left(\frac{n\pi y}{d}\right) \quad n:\ \text{odd},$$

$$\phi_n(y) = \sqrt{\frac{2}{d}} \sin\left(\frac{n\pi y}{d}\right) \quad n:\ \text{even},$$ (2.20c)

respectively. Here, d is the size of the square. The energy $E_{l,n}$ of the electrons in the 2D square well is given by the sum of the quantized energies of electrons moving along the x- and y-directions. Namely, the energy $E_{l,n}$ is expressed by the following equation for the 2D square well with infinite confinement potentials:

$$E_{l,n} = \frac{\hbar^2}{2m}\left\{\left(\frac{l\pi}{d}\right)^2 + \left(\frac{n\pi}{d}\right)^2\right\}.$$ (2.20d)

State-of-the-art modern nanoscience and technology can be used to actually construct nanoscale 2D square and rectangular quantum wells of arbitrary size by manipulating individual atoms on atomically flat substrate surfaces by an STM. A demonstration is presented in Figure 2.15 [4]. In this case, first, a small number of Mn atoms are deposited on an atomically clean and flat Ag(111) substrate surface. Then, the neighboring 28 Mn atoms are selected from the deposited Mn atoms and manipulated by "drag and drop" using the tip of an STM to construct an aligned

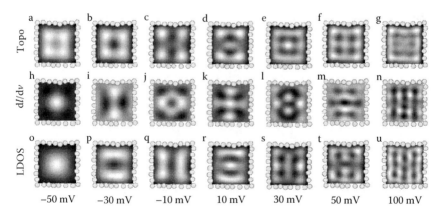

Figure 2.15 An example of 2D nanosquare formation using the atom manipulation in STM. (Reprinted from J. Kliewer, R. Brendt, and S. Crampin, *New J. Phys.* 3, 2001, 22. With permission.)

structure consisting of a 9×10 nm rectangle on the surface. Since the STM tip can be used to put the atoms at any desired position on the surface with atomic resolution, the atom corral of any desired shape and size can be constructed by this method.

Now, the crystalline Ag substrate has a 2D—free-electron-like—electronic band structure, which is localized at its (111) surface. The adatoms scatter the 2D free electrons at the surface. Thus, the array of the adatoms behaves as potential energy walls for the surface electrons. Utilizing these properties, the Mn atom corral is used to confine the 2D free electrons in the surface.

The atomic resolution of an STM can be used not only for constructing an atom corral, but also for characterization of the structure and electron states of the atom corral. In an STM, surface undulations and, in some cases, the spatial distribution of electrons with energies equal to the applied bias voltage between the tip and the surface can be mapped in conventional topographic image mode (the upper column labeled "topo" in the figure). Details about STMs are given in an aside later in this chapter. In the case of Mn on Ag substrates, the image of individual Mn atoms of the corral appears as protrusions on the atomically flat Ag(111) surface as shown in Figure 2.15a. Inside the corral, the spatial distribution of the confined electrons is also represented in the topographic STM images. The bias voltage, namely the energy of the imaged electrons with respect to the Fermi level, is shown at the bottom of each image in the figure.

The size along the x- and y-axes was very close in the present 28 Mn atom corral. Therefore, we examine its QWSs using the equation for the 2D square. From Equation 2.20, the lowest QWS occurs for $(l, n) = (1, 1)$. In this case, the wave functions $\phi_l(x)$ and $\phi_n(y)$ have a peak at the center in both the x- and y-directions. As a result, the electron density $(|\psi(x, y)|^2)$ is expected to be a maximum at the center of the square. Similarly, the second lowest QWSs appear for $(l, n) = (2, 1)$ and $(1, 2)$. The wave function $\phi_l(x)$ has a peak and a valley along the x-direction, while $\phi_n(y)$ has one at the center along the y-direction in the former case. At the superposition of $|\phi_l(x)|^2$ and $|\phi_n(y)|^2$, the electron density $|\psi(x, y)|^2$ is expected to have two peaks splitting along the x-direction. In the latter case $(l, n) = (1, 2)$, the electron density has two peaks splitting along the y-direction. Furthermore, the third lowest QWS with $(l, n) = (2, 2)$ is expected to have $2 \times 2 = 4$ peaks along both the x- and y-directions as shown in Figure 2.15c. The higher states with $(l, n) = (3, 3)$ and $(4, 4)$ are expected to have the electron spatial distribution $|\psi(x, y)|^2$ like Figures 2.15f and g, respectively. The STM images clearly reproduced the theoretically expected results in which QWSs with higher energies appear sequentially with increasing bias voltage, which is equal to the energy of electrons detected to construct the STM images.

2.3.2   2D cylindrical wells

Next, examine quantum confinement in 2D cylindrical potential wells. For simplicity, we assume that the confinement potential is infinitely high, which is the same as the treatment in the 2D square well case. Since the 2D cylindrical well system has a circular symmetry around the z-axis, it is convenient to describe the Schrödinger equation for electron motion in the x–y-plane using polar coordinates, that is, $(x, y) \Leftrightarrow (r, \theta)$ in the present case as

$$\left\{ -\frac{\hbar^2}{2m}\frac{\partial^2}{\partial r^2} + \frac{1}{r}\frac{\partial}{\partial r} + \frac{1}{r^2}\frac{\partial^2}{\partial \theta^2} \right\} \phi_{x,y}(r, \theta) = E_{x,y}\phi_{x,y}(r, \theta). \qquad (2.21a)$$

Now, the motion of free electron along the z-axis is described by the following equation:

$$-\frac{\hbar^2}{2m}\frac{\partial^2}{\partial z^2}\phi_z(z) = E_z\phi_z(z). \qquad (2.21b)$$

In the x–y-plane, electrons move freely as described by Equation 2.21a. However, in the present case of the cylindrical well, their motion is restricted to inside of the cylinder due to the infinite potential wall at $r = a$ (a is the radius of the nanowire). This situation is mathematically represented by the boundary condition $\phi_{x,y}(a, \theta) = 0$ for Equation 2.21a. In addition, one more boundary condition appears for the wave function $\phi_{x,y}(r, \theta)$ due to the rotational symmetry of the 2D cylindrical system. The system dictates the rotational symmetry as being $\phi_{x,y}(r, \theta) = \phi_{x,y}(r, \theta + 2\pi)$ (i.e., the system is equivalent for the rotation $\theta = 2\pi$). Thus, the wave function should take the form $\phi_{x,y}(r, \theta) = \phi_R(r) \exp(il\theta)$. Here, l is the so-called angular momentum quantum number—an integer of the angular momentum quantum number l—and the electron motion is quantized in the rotational direction. Substituting the expression $\phi_{x,y}(r, \theta) = \phi_R(r) \exp(il\theta)$ in Equation 2.21a, we obtain the following Schrödinger equation for the radial part of wave function $\phi_R(r)$:

$$\left\{ -\frac{\hbar^2}{2m}\left(\frac{d^2}{dr^2} + \frac{1}{r}\frac{d}{dr} \right) + \frac{\hbar^2 l^2}{2mr^2} \right\} \phi_R(r) = E_{x,y}\phi_R(r). \qquad (2.22)$$

Putting $E_{x,y} = \hbar^2 k_\parallel^2/2m$ and $z = k_\parallel r$ in Equation 2.22, the above equation is modified as

$$\frac{d^2\phi_R(z)}{dz^2} + \frac{1}{z}\frac{d\phi_R(z)}{dz} + \left(1 - \frac{l^2}{z^2} \right)\phi_R(z) = 0. \qquad (2.23)$$

This is mathematically equivalent to the *l*th Bessel differential equation, the solution of which is given by a linear combination of the *l*th Bessel and Neuman function, $J_l(z)$ and $N_l(z)$. However, in our case, we omit any $N_l(z)$ terms from the solution since $N_l(z)$ diverges at $z = 0$ and is unphysical. After all, the physically meaningful solution of Equation 2.21a is given as

$$\phi_{x,y}(r, \theta) \propto J_l(k_\| r)\exp(il\theta) \qquad (2.24)$$

Here, the Bessel function $J_l(z)$ is an oscillatory function. It behaves asymptotically $(\approx\sqrt{2/\pi z}\ \cos(z - (l/2)\pi - (\pi/4)))$ for large z and crosses zero at $z = j_n$, where $n = 0, 1, 2, \dots$, as shown in Figure 2.16. Numerical values of j_n are shown for several values of l in the table. Using the tabulated values of j_n, the radial boundary condition $\phi_{x,y}(a, \theta) = 0$ is expressed as $k_\| a = j_n$. The condition $k_\| a = j_n$ poses a restriction on $k_\|$, and the wavevector $k_\|$ is quantized. Finally, the wave function $\phi_{x,y}(r, \theta)$ is fully quantized by the combination of two quantum numbers (n, l). The corresponding energy of the confined electron $E_{x,y} = \hbar^2 j_n^2 / 2ma^2$ is also quantized by the quantum numbers (n, l) in a discrete manner.

In a similar manner to the previous example of the rectangular atom corral, a nanoscale cylindrical well can be constructed by manipulating atoms on atomically flat surfaces. Figure 2.17 shows the STM images before (the upper panel) and after (the lower panel) construction of a cylindrical well with 32 Ag atoms on an Ag(111) surface [5]. The diameter of the corral is 32 nm. Construction and measurement are carried out using a cryogenic STM at 6 K. STM scan images show the 32 protrusions corresponding to atoms that were manipulated to form the circular line. In addition, the concentric wavy patterns in the corral represent the peaks of the special distribution, $|\psi(r, \theta)|^2$, of confined electrons. In the completed corral, five concentric circles are visible, although their angular distributions are averaged to be indistinguishable. Along the radial direction, the first peak appears at an off-centered position in the corral. Referring to the figure of the Bessel function and its zero

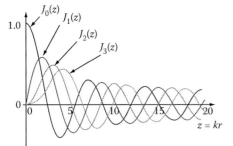

	$l = 0$	$l = 1$	$l = 2$	$l = 3$
$n = 1$	2.4	3.8	5.2	6.4
$n = 2$	5.6	7.1	8.4	9.8
$n = 3$	8.7	10.2	11.6	13.0
$n = 4$	11.8	13.3	14.8	16.2

Figure 2.16 Bessel function and its zero cross points.

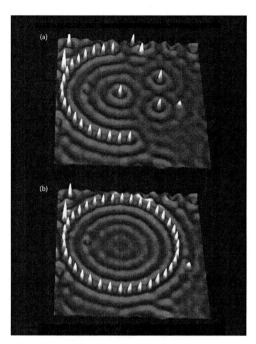

Figure 2.17 **(See color insert following page 148.)** Construction of a 2D cylindrical well using STM (a) during the construction and (b) the completed cylindrical well. (Reprinted from S. Hla, K. Braum, and K.H. Rieder, *Phys. Rev. B* 67, 2003, 201402. With permission.)

cross-point table, we notice that the first radial peak at the off-centered position means that the observed quantized state corresponds to $l \pi 0$ quantum states. This is because the peak in $\phi_R(r)$ appears at the center only for the state with $l = 0$, as indicated in the figure of the Bessel function. For the state $l \neq 0$, the first peak appears at an off-centered position, and the number of peaks in the radial direction (equals the number of concentric patterns in the corral) corresponds to the quantum number n, as illustrated in the figure and table. Therefore, the STM image in the lower panel suggests that the observed $|\psi(r, \theta)|^2$ is for the cylindrical QWS with $l \neq 0$ and $n = 5$.

2.3.3   Shape effect on the quantized states

Next, we examine the effect of shape on the 2D quantized states by comparing the QWSs in a 2D square and cylindrical wells of the same size. As an example, we consider the QWSs in a 2D square with a size $d = 10$ nm and those in a 2D cylindrical well with a diameter $d = 10$ nm (1×10^{-8} m). The square well of $d = 10$ nm encloses the cylindrical well of

$d = 10$ nm (i.e., the radius $a = 5$ nm) inside it, as illustrated in the lower panel in Figure 2.18. Keeping in mind that $m = 9.1 \times 10^{-31}$ kg, $\hbar = 1.0 \times 10^{-34}$ J s, and 1 eV $= 1.6 \times 10^{-19}$ J, we can numerically calculate the QWS energy $E_{l,n} = \hbar^2/2m\{(l\pi/d)^2 + (n\pi/d)^2\}$ for the 2D square well and $E_{x,y} = \hbar^2 j_n^2/2ma^2$ for the 2D cylindrical well. The calculated quantized state energies are shown on the left and right sides of Figure 2.18, respectively.

The figure shows that the quantized state energy is generally higher in the cylindrical well than in the square well. This is because the confinement area in the cylindrical well is smaller than that in the square well as illustrated in Figure 2.18. As surmised in the simple 1D quantum well (Equation 2.19), if electrons are confined into smaller areas, then the quantized state energy levels are pushed higher up. Furthermore, there are several degenerate states in the square well. These states are easily split into states with small energy difference, when the shape changes from high-symmetric square. The point is that the results in Figure 2.18 demonstrate the possibility of tailoring the electronic states of nanomaterials by controlling their shape artificially. As discussed later, a wide energy separation is favorable for the quantized states to maintain their quantum nature at high temperatures against thermal fluctuations. From this viewpoint, making materials smaller in size and with highly symmetrical shapes is an important strategy in designing nanomaterials exhibiting robust quantum-mechanical properties.

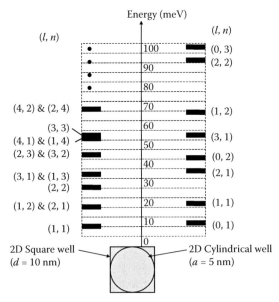

Figure 2.18 Numerical results of the quantized states in the 2D square and cylindrical wells.

2.3.4 Finite potential wells

So far, we have assumed that the confinement potential in quantum wells is infinite. However, in practice, this potential is finite in realistic nano-systems. The existence of a finite confinement potential modifies the results obtained for the infinite potential cases by (i) lowering the quantum energy levels and (ii) electron spillage from the quantum well. We discuss these two effects in detail for the simplest 1D quantum well case.

Here, we assume that the quantum well has a potential barrier of height V_c at $x = \pm d/2$ (where d is the well width). The potential for the electron is

$$V(x) = \begin{cases} V_c & \text{at } |x| \geq \pm\dfrac{d}{2} \\ 0 & \text{at } \dfrac{d}{2} \geq x \geq -\dfrac{d}{2}. \end{cases} \quad (2.25)$$

In the quantum well, electrons with energies E ($<V_c$) are confined, and have wave functions given by

$$\psi(x) = A\exp(ikx) + B\exp(-ikx) \quad \text{for } \frac{d}{2} \geq x \geq -\frac{d}{2}. \quad (2.26a)$$

However, the electron wave functions decay rapidly outside the well as

$$C\exp(-\kappa x) \quad \text{for } x \geq \frac{d}{2}, \quad (2.26b)$$

$$D\exp(+\kappa x) \quad \text{for } x \leq -\frac{d}{2}. \quad (2.26c)$$

Here, $k = \sqrt{2mE}/\hbar$ and $\kappa = \sqrt{2m(V_c - E)}/\hbar$. Since $V(x)$ is symmetric around $x = 0$, $|\psi(x)|^2 = |\psi(-x)|^2$ should be valid in this system. Thus, the wave function should satisfy the following relationships for the inversion of the coordinate x:

$$\psi(x) = \psi(-x) \quad \text{(symmetric)},$$
$$\text{or } \psi(x) = -\psi(-x) \quad \text{(antisymmetric)}. \quad (2.27)$$

These relationships holds for $B = \pm A$ and $D = \pm C$ (+: symmetric, −: antisymmetric). Utilizing these relationships between the coefficients A, B, C, and D, the boundary conditions [continuity of $\psi(x)$ and $d\psi(x)/dx$ at $x = \pm d/2$] yield following equations:

$$\tan\left(\frac{kd}{2}\right) = \sqrt{\frac{2mV_c}{\hbar^2 k^2} - 1} \quad \text{(symmetric)},$$
$$-\cot\left(\frac{kd}{2}\right) = \sqrt{\frac{2mV_c}{\hbar^2 k^2} - 1} \quad \text{(antisymmetric)}. \quad (2.28)$$

Solving the above equations graphically, the energy levels of the QWSs are obtained for a finite confinement potential well system.

Figure 2.19 shows two examples of the graphical solutions of Equation 2.28. The results for a system with $V_c = 5$ eV, $d = 1$ nm are shown in the upper panel. The lower panel stands for results for a system with $V_c = 0.1$ eV, $d = 10$ nm. From these graphs, we obtain the specific k for the quantized states. The k is translated to the eigen energy of the quantized states by using the relationship $k = \sqrt{2mE}/\hbar$. The V and d in these examples are typical for nanostructures utilizing clusters (in the former case) and semiconductor superlattices (in the latter case). The numerical values of the quantized states, which are obtained from the graphs in Figure 2.19, are listed in Table 2.1. As a reference, the results for the infinite confinement potential are also given in parentheses in the table. Figure 2.20 shows the numerically calculated wave function of the $n = 2$ QWS for $V_c = 5$ eV and ∞.

The results in the table show that, compared with those in infinite wells, wavenumbers and energy levels decrease in the finite potential barrier well. This is due to the spillage of the confined electrons from the finite potential barrier. The $|\psi(x)|^2$ does not necessarily fall to zero at the boundaries of the well. The probability function is allowed to spill out into the potential barrier region with an exponential decay in the finite well as shown in Figure 2.20, where the wave function of the $n = 2$ quantized state spills into the potential barrier regions for $V_c = 5$ eV, whereas it is totally confined in the well for $V_c = \infty$. As shown in the figure, the wave functions of the confined electrons elongate slightly toward the outside of the well in the finite well case. This results in an increase in the wavelength of the

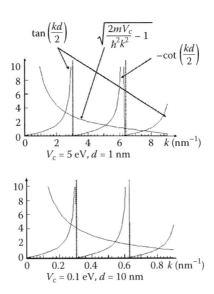

Figure 2.19 Graphical solution of the quantized state's k in 1D finite potential wells.

Table 2.1 Numerical Results of k and Energy Level of Quantized States in Finite Potential Wells

		k (nm^{-1})	κ (nm^{-1})	E (eV)
$n = 1$	$V_c = 5$ eV, $d = 1$ nm	2.6	11	0.25
	$(V_c = \infty, d = 1$ nm)	(3.1)	(0)	(0.36)
$n = 2$	$V_c = 5$ eV, $d = 1$ nm	5.1	10	0.98
	$(V_c = \infty, d = 1$ nm)	(6.2)	(0)	(1.45)
$n = 3$	$V_c = 5$ eV, $d = 1$ nm	7.5	8.7	2.1
	$(V_c = \infty, d = 1$ nm)	(9.4)	(0)	(3.3)
$n = 1$	$V_c = 0.1$ eV, $d = 10$ nm	0.26	1.6	0.0025
	$(V_c = \infty, d = 10$ nm)	(0.26)	(0)	(0.0025)
$n = 2$	$V_c = 0.1$ eV, $d = 10$ nm	0.54	1.5	0.011
	$(V_c = \infty, d = 10$ nm)	(0.62)	(0)	(0.014)
$n = 3$	$V_c = 0.1$ eV, $d = 10$ nm	0.81	1.4	0.024
	$(V_c = \infty, d = 10$ nm)	(0.94)	(0)	(0.033)

confined electrons (equal to a decrease in wavenumber vector k). A reduction in the value of k lowers the energy levels of quantum states for the free-electron-like dispersion $E = \hbar^2/2mk^2$, which we assumed here.

The numerical results in the table tell us that there is a considerable reduction in the quantized state energy in narrow quantum wells. For example, the energy is lowered by ~30% for every QWS in 1 nm quantum wells with a 5 eV potential barrier. It is also noteworthy that the decay length of $\psi(x)$ in the finite potential wall (i.e., $1/\kappa$) is of the order 0.1 nm for $V = 5$ eV, whereas it is 1 nm for $V = 0.1$ eV.

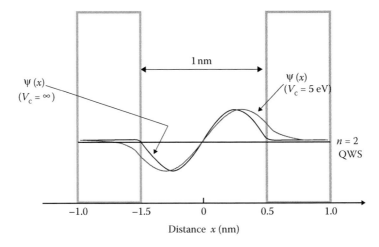

Figure 2.20 1D quantum well system with a finite potential barrier.

AN ASIDE: ORIGIN OF CONFINEMENT POTENTIALS

In the preceding discussion, we discussed the confinement potential barrier *a priori* without considering its origin. In practice, potential barriers can be produced by electric potentials induced externally by electrodes, by band offsets at interfaces between heterostructures of, for example, semiconductors, and work function difference between materials (Figure 2.21).

The electrode-induced potential is advantageous in that the potential height can be controlled freely by the bias voltage applied to the electrodes. The spatial distribution of the potential area is also controllable through the spreading of equipotential lines around the electrode by increasing the applied bias voltage. A typical example is the split gate structure, in which two metal gates are defined with a narrow space in a material using fabrication techniques such as the electron beam lithography. The application of a negative bias voltage on both gates causes the repulsive electrostatic potential to extend around the gates. By increasing the negative bias voltage, the space between the repulsive potential can be reduced to the nanometer scale. In this case, the potential height is fundamentally of the same order as the applied bias voltage.

Band offsets are also used to define potential wells. For example, molecular beam epitaxy is used to fabricate multilayered structures in which nanometer-thick films of different materials are stacked. A typical example is the GaAs/AlGaAs superlattice, where both GaAs and AlGaAs have a zinc-blend crystal structure. In addition, the lattice constants of these two compound semiconductors match almost perfectly. This enables the fabrication

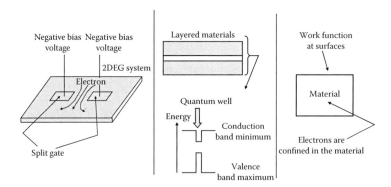

Figure 2.21 Confinement of electrons; left: split gate; middle: layered structure; right: work function.

of smoothly connected layered materials. However, the bandgap of GaAs is smaller than that of AlGaAs. Therefore, a potential well appears in the GaAs layer, which is sandwiched by AlGaAs layers. The potential height is determined by the relative alignment of the energies of the conduction band minima of the two semiconductors. It is typically in the order of 0.1 eV, and the electrons are confined to move along a perpendicular plane to the interface of the sandwiched GaAs layer and form the subband electronic structure.

Without employing such artificial procedures, electrons are naturally confined in the bulk and cannot spontaneously escape from their mother materials based on the work functions of the materials. Using this intrinsic nature, we can confine electrons in nanospace just by reducing the size of materials. This spontaneous confinement is natural, intrinsic, and useful in many aspects of nanoscience and technology. In this respect, we briefly discuss its origin in the following.

The question to consider here is: what causes the valence electrons to be confined in their mother (host) materials? A simple and intuitive answer is the existence of the electrostatic Coulomb potential. Atoms have an electronic structure of closed-shell + outermost electron(s). In solid materials, atoms eject their outermost electron(s) in the form of valence electrons. These ejected valence electrons move freely, while the positively charged ion cores are static at the atomic position in the solid. However, in principle, the charge neutrality [i.e., $\rho_+(r) = \rho(r)$] should hold at any location in the solid. Here, we assume the spatial density of the positively charged ion cores and the negatively charged valence electrons to be $\rho_+(r)$ and $\rho(r)$, respectively. But, it is not valid at the surfaces of solids. In a solid, the positively charged ion cores are aligned regularly. This is reflected by the constant positive charge density $\rho_+(r)$ in the bulk. However, there are no ion cores outside the solid surfaces. Thus, $\rho_+(r)$ is abruptly truncated stepwise, to zero at a surface, as illustrated in Figure 2.22 [6]. However, the electron density distribution $\rho(r)$ cannot follow this abrupt change and oscillates near the surface (the so-called Friedel oscillation). As a result, a net charge $[\rho(r) - \rho_+(r) \neq 0]$ appears at the surface, which results in an electrostatic Coulomb potential:

$$\phi(\mathbf{r}) = \int \frac{\rho(\mathbf{r}) - \rho_+(\mathbf{r})}{|\mathbf{r} - \mathbf{r}'|} \, d\mathbf{r}'. \tag{2.29}$$

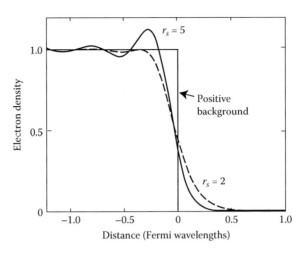

Figure 2.22 Spatial distribution of positive and negative charge density at a surface. (Reprinted from N. Lang and W. Kohn, *Phys. Rev. B* 1, 1970, 455. With permission.)

For a model in which the positive change distribution is regarded to be constant in a solid (the so-called jellium model), a density functional theory calculation reveals the charge distribution to be as shown in Figure 2.22. The net difference between $\rho_+(\mathbf{r})$ and $\rho(\mathbf{r})$ is positive inside the bulk of the solid, whereas it is negative outside the surface. This net charge distribution generates a dipole layer at the surface and stabilizes electrons in the solid by lowering the electrostatic potential inside the solid by $\Delta\phi$.

In addition to the dipole field, the exchange-correlation effect $|V_{xc}|$ (a many-body effect in the interaction between valence electrons) also lowers the energy of electrons. Thus, the potential for the valence electrons is lowered totally $\Delta\phi + |V_{xc}|$ with respect to the potential outside the solid surface (vacuum). Meanwhile, the valence electrons move with kinetic energies $E_{kin} = \hbar^2 k^2/2m$ in the potential of the bulk of the solid. Now, assuming that the maximum kinetic energy of electrons is given by $E_F = \hbar^2 k_F^2/2m$ at the Fermi level, a minimum amount of work given by $\Delta\phi + |V_{xc}| - (\hbar^2 k_F^2)/2m$ is necessary to remove a valence electron from a solid to the vacuum, as illustrated in Figure 2.23. Finally, valence electrons are stabilized in a solid by the potential barrier equal to this work. This energy barrier is called the *work function*. For normal metals, $\Delta\phi$, $|V_{xc}|$, and $E_F = \hbar^2 k_F^2/2m$ are approximately 7, 10, and 13 eV, respectively [7]. This implies that the work function is around 4 eV. Actually, the measured work function for normal metals is around 4–5 eV.

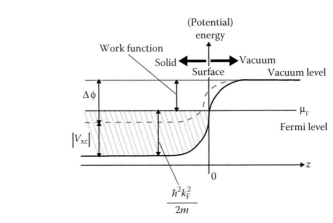

Figure 2.23 Energy diagram of the work function at surfaces.

2.3.5 Band dispersion effect

In the previous sections on quantum confinement, we assumed *a priori* that the system exhibits free-electron-like dispersion with an effective mass of *m* given by

$$E(k) = \frac{\hbar^2}{2m} k^2. \tag{2.30}$$

This is because in many cases, the band dispersion $E(k)$ can be approximated well by the free-electron-like dispersion. However, the validity of free-electron-like dispersion depends on the curvature of the dispersion and the energy range of interest. In some cases, the real band dispersion cannot be approximated with the required accuracy by free-electron dispersion. We must take intrinsic band dispersion into account while calculating quantum confinement in such cases.

For example, here we consider quantization in a 1D quantum well with a tight-binding-model-like band dispersion as

$$E(k) = \alpha + 2\gamma \cos(ka). \tag{2.31}$$

A simple way to take dispersion into account is use of the Bohr–Sommerfeld condition for quantization. This condition is sometimes called the phase accumulation rule. The Bohr–Sommerfeld condition can be described as

$$2k(E)d + \delta_R(E) + \delta_L(E) = 2\pi n, \tag{2.32}$$

for a 1D quantum well, with width d and energy dispersion of $E(k)$. Here, $\delta_R(E)$ and $\delta_L(E)$ represent the phase shift when an electron with energy E bounces off the right and left boundaries of a potential well, respectively (Figure 2.24). Also, $k(E)$ is the reciprocal function of the dispersion $E(k)$. In a well of width d, Equation 2.32 can be satisfied only for specific values of k. The quantized states are obtained by substituting the values into the dispersion relationship, $E(k)$.

Equation 2.32 describes quantized conditions for the following reason. In a 1D quantum well, electrons move freely and make round trips many times by repeated scattering (bouncing) at the potential walls on both sides. The free electrons have wave functions defined by $\psi(z) \propto \exp(ikz)$. Typically, in a round trip starting from the point A (Figure 2.24), an electron moves freely from A to B, bounces at the right-side potential barrier B, and then travels freely from B to C, bounces off the left potential barrier C, and finally goes back to the starting point A. The electron travels a distance $2d$ during one round trip. As a result, the wave function $\psi(z) \propto \exp(ikz)$ accumulates a phase $2kd$. In addition, bouncing at the right- and left-side potential barriers causes a change of phase given by δ_R and δ_L, respectively. In total, the electron gains a phase of $2k(E)d + \delta_R(E) + \delta_L(E)$ in one round trip. Here, there is constructive interference between the returned electron wave function and the original one, and a standing wave (quantized state) is maintained even after many times repetitions of the round trip by electrons in the 1D quantum well, if $2k(E)d + \delta_R(E) + \delta_L(E)$ is a multiple of 2π, that is, $2\pi n$, where n is an integer. However, the returned electron waves interfere destructively if the condition, Equation 2.32, is not satisfied. In this case, the phase differences are accumulated randomly by repeated round trips. After repetition of many round trips, the interference becomes totally destructive, and no electron states survive. Thus, the phase accumulation rule of Equation 2.32 gives the conditions for quantization.

Especially, in the case of a hard wall (i.e., infinite confinement potential wall), $\delta_R(E) + \delta_L(E) = 2\pi$. Therefore, the Bohr–Sommerfeld condition is

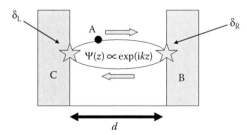

Figure 2.24 Phase accumulation during the round trip of electrons confined in a 1D quantum well.

described as $k(E) = \pi/dn$. It is noteworthy that this equation is highly consistent with the results for free-electron dispersion described by Equations 2.18a and 2.18b. However, the Bohr–Sommerfeld condition is more flexible in that it is easily applicable to the finite confinement potential $[\delta_R(E) + \delta_L(E) \neq 2\pi]$ cases, and also to the nonparabolic dispersion cases.

Here, we use the phase accumulation rule of Equation 2.32 to estimate the quantized states in a 1D quantum well in which electrons exhibit nonparabolic dispersion defined by Equation 2.31. In this calculation, we assume that electrons are confined in a well of width $d = 2$ nm and that the confinement potential is infinite. We put a set of parameters $\alpha = -0.60$ eV, $\gamma = -0.22$ eV, and $a = 0.5$ nm—typical for conventional materials—in the dispersion Equation 2.31. As illustrated in Figure 2.25, this dispersion (in the right panel) differs from that of parabolic dispersion. However, it can be approximated quite well at around its bottom by assuming parabolic dispersion. By fitting this dispersion with a parabolic one, we obtain the approximate parabolic dispersion shown in the left panel in Figure 2.25. For both dispersions, the energy levels of the QWSs are obtained numerically by applying the Bohr–Sommerfeld condition. The results are indicated by horizontal lines in the figure. The numerical results are also indicated in Table 2.2. Since the parabolic dispersion was a good approximation of the correct at around the bottom, the energy levels of the lowest and the second lowest QWSs appear at similar energies in both cases, as shown in Figure 2.25 and Table 2.2. However, the discrepancy in the potential landscape becomes larger

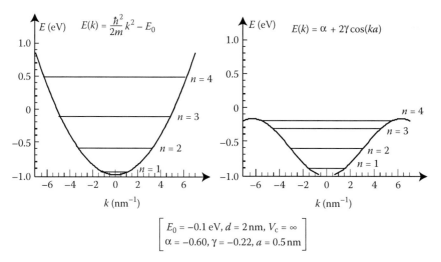

Figure 2.25 Numerically calculated 1D QWSs with a parabolic (left) and a nonparabolic (right) dispersion.

Table 2.2 Numerically Calculated 1D QWSs with a Parabolic a and Nonparabolic Dispersion

	k (nm^{-1})	$E(k) = (\hbar^2/2m)k^2 + E_0$ (eV)	$E(k) = \alpha + 2\gamma \cos(ka)$ (eV)
$n = 1$	1.57	−0.90	−0.89
$n = 2$	3.14	−0.62	−0.59
$n = 3$	4.71	−0.16	−0.29
$n = 4$	6.28	0.49	−0.17

Notes: $E_0 = -0.1$ eV, $d = 2$ nm, $V_c = \infty$, $\alpha = -0.60$, $\gamma = -0.22$, $a = 0.5$ nm.

at higher energies between the correct and the approximate parabolic dispersions, as illustrated in the figure. This results in an inevitably larger error in the parabola approximation (see the figure and the table). This demonstrates the importance of taking the correct dispersion into account in the numerical design of QWSs.

2.4 Edge (surface)-localized states

The surface-to-volume ratio of materials increases dramatically when their size is reduced. This is easily understood by considering the size-dependent change of the surface and volume of a spherically shaped material. The surface area and the bulk volume of a spherical material of radius r are $S = 4\pi r^2$ and $V = 4/3\pi r^3$, respectively. Therefore, the S/V ratio increases with decrease in the size r as r^{-1}. This means that the smaller the size, the larger the contribution of surfaces to the properties of materials. The edges of materials (surfaces) play a dominant role in governing the physical properties of extremely small materials, where most of the atoms sit, not in the bulk, but at the surfaces as illustrated in Figure 2.26.

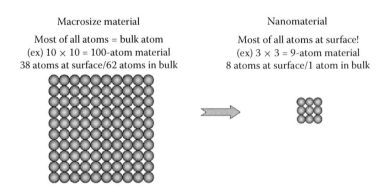

Macrosize material

Most of all atoms = bulk atom
(ex) 10 × 10 = 100-atom material
38 atoms at surface/62 atoms in bulk

Nanomaterial

Most of all atoms at surface!
(ex) 3 × 3 = 9-atom material
8 atoms at surface/1 atom in bulk

Figure 2.26 Size-dependent change of the surface/volume ratio.

In many cases, the surfaces of materials possess exotic localized electronic states. However, the role of the surface-localized electronic states can be neglected in macroscopic-sized materials and most of the properties can be understood in terms of the bulk electron states. On the other hand, electron states localized at the surface play an important role in nanomaterials. Especially, this is the case when the size of the material is ultimately small. The exotic nature of the surface-localized electron states can be understood intuitively, for example, by considering the surface of crystalline group IV materials (C, Si, Ge, etc.), which have a diamond-type crystalline structure and many are semiconductors. A group IV atom has four sp^3-hybridized orbitals. The diamond structure is constructed via the formation of interatomic chemical bonds between neighboring atoms with sp^3-hybridized orbitals. Here, the formation of an interatomic bond makes the two sp^3-hybridized orbitals on the atoms of both sides of the bond split into the bonding and the antibonding states as illustrated in the left panel in Figure 2.27. The bonding (antibonding) state is energetically lower (higher) than the sp^3-hybridized orbital state. The two valence electrons at the sp^3-hybridized orbitals are now accommodated in the energetically lower bonding state in the interatomic bond. The bonding state is fully occupied, whereas the antibonding state is completely empty. This lowers the electron energy and makes the interatomic bond strong and energetically stable.

However, every atom has neighboring atoms at all the ends of its four sp^3-hybridized orbitals in the bulk. The formation of interatomic bonds and the subsequent splitting of the sp^3-hybridized orbitals into bonding and antibonding states occur everywhere. It means that both the occupied bonding states and the empty antibonding states are highly

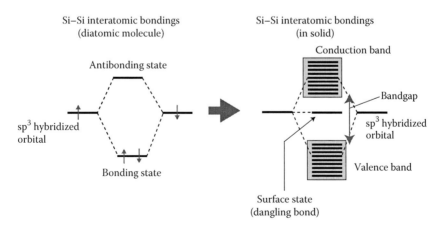

Figure 2.27 Electronic states of bulk- and surface-localized electronics states of group IV materials.

degenerated. But, the degeneracy is actually lifted in materials by the crystal potential. As a result, the highly degenerated states split into many states around the original energy position of the bonding and anti-bonding states with a very small energy spacing. The split states around the bonding and antibonding states form the valence and conduction bands, respectively. A bandgap also appears between the valence and conduction bands, since the bonding and antibonding states were originally separated energetically. Following this scenario, group IV materials exhibit semiconducting electron band structures as illustrated in the right panel in Figure 2.27.

However, the atoms at the surfaces have no neighboring atoms on their vacuum side. As a result, the unpaired bond with the intrinsic nature of sp^3-hybridized orbitals remains so as to intrude toward the vacuum side at surfaces. These unpaired bonds are called *dangling bonds*. Since dangling bonds have no bonding pairs, they keep their original sp^3-hybridized bond nature, the energy of which is at the center of the bandgap as shown in the right panel in Figure 2.27. In bulk materials, the electrons are forbidden to occupy electron states within the bandgap. In this respect, surface dangling bonds are recognized as exotic electron states, which appear to be localized at the surfaces of materials.

Dangling bonds cost energy because electrons in the dangling bonds cannot be accommodated into the energetically lower-lying bonding states. These dangling bonds dominate the electron structure in extremely small-sized materials. More generally, the surface is energy costing in a wide variety of materials. In some cases, the energetically unfavorable surface nature works as a driving force to change the shape of nano-materials. For example, the material takes a spherical shape to minimize the surface energy, if the energy cost at surfaces is isotropic. This is because the sphere is the shape with the smallest surface area under a given volume. In the case when energy cost is anisotropic and changes depending on the surfaces crystallographic orientations, the nanomaterials take the polyhedron shape with specific surfaces of low surface energy.

The above discussion tells us intuitively of the existence of exotic surface-localized electronic states. For example, we saw that the dangling bond is exotic in that it appears in the bulk-forbidden bandgap. However, it is not clear whether this electron state has a localized nature at surfaces. In the following, we show the localized nature of states by extending the conventional two-band model description of the bulk band structure so as to include imaginary wavevectors.

In the conventional two-band model, electron states in crystalline bulk materials are described by the sum of two electron waves with wavevectors k and $k + g$ given by

$$\psi(x) = A(0)\, e^{ikx} + A(g)\, e^{i(k-g)x}. \qquad (2.33)$$

Here, g is the reciprocal lattice vector of the crystalline material. Substituting this wave function into Schrödinger equation and solving the eigenstate equation, we obtain the following bulk electronic band dispersion relationship:

$$E(k) = \frac{1}{2}\frac{\hbar^2}{2m}\{k^2 + (k+g)^2\} \pm \frac{1}{2}\sqrt{\left(\frac{\hbar^2}{2m}\right)^2 \{k^2 - (k+g)^2\}^2 + 4V_g^2}, \quad (2.34)$$

where V_g is the Fourier component of the crystalline potential for the reciprocal lattice vector g. In conventional theory of bulk band structure, the wavevector k is recognized as a real number *a priori*. The real k means that the electron wave function of Equation 2.33 delocalizes completely in the bulk region. For a real wavenumber k, Equation 2.34 yields two energy states $E(k)$ in the energy range corresponding to the conduction (+) and valence (–) band as illustrated in Figure 2.28. However, Equation 2.34 does not have a solution for an energy E in the bandgap for any real wavenumber k. It means that electrons with energy E in the bandgap cannot be entirely delocalized in the bulk. Thus, an electron state with energy E in the bandgap is forbidden as the bulk electronic state. In this manner, the two-band model succeeds in describing the two (valence and conduction) bulk electron bands separated by a bandgap.

However, if the wavevector k is taken to include an imaginary part $k = g/2 - i\kappa$, then an electron state of energy $E(k)$ is allowed to exist in the bandgap as illustrated in Figure 2.28. More specifically, let us consider the situation in that a bulk material occupies the region $x \le 0$ (i.e., the surface is located at $x = 0$ and the region $x > 0$ is vacuum) as shown in Figure 2.29.

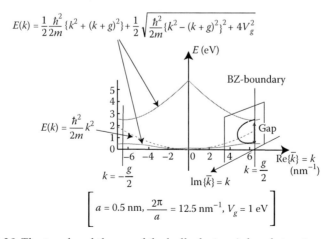

Figure 2.28 The two-band theory of the bulk electronic band structure.

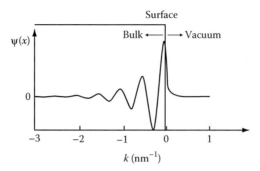

Figure 2.29 The spatial distribution of surface-localized electronic state.

Substitution of the imaginary wavenumber into Equation 2.33 shows that the oscillation of the electron wave function $\psi(x)$ dampens with distance from the surface as $\psi(x) \propto e^{\kappa x}$, as shown for $x \leq 0$ in Figure 2.29. The electron wave function $\phi(x)$ also decays outside the material ($x > 0$). Here, the wave functions $\psi(x)$ and $\phi(x)$ must be continuous at the surface $x = 0$ as shown in the figure. The matching conditions determine the energy $E(k)$ of the wave functions in the bandgap. The resulting wave function is localized at the surface as shown in Figure 2.29. These kind of surface-localized states are known as Shockley states.

Apart from the mechanisms discussed above, local modulation of crystalline potentials also causes localization of electron states at surfaces. States localized at surfaces due to these physical mechanisms are called Tamm states. This discussion shows that electrons have a strong tendency to form exotic surface-localized states via both the Shockley and Tamm state mechanisms. These surface-localized states play an important role in ultrasmall nanomaterials.

2.5 Charging effect

Electrons have an elementary charge of 1.64×10^{-19} C. This was measured experimentally in 1909 by Millikan and Fletcher for the first time. The key to the success of this experiment was the use of sprayed oil droplets, which were small enough to allow adhesion of few electrons. They determined the charge of an electron by measuring the electric field necessary to stop free falling droplets due to gravitational forces. They found that the charge of droplets was always multiples of an elemental value and determined the elementary charge of an electron.

In the Millikan and Fletcher experiments, small oil droplets were necessary to reduce the number of attached electrons to be small enough so that they could be counted one by one. In addition, small-sized droplets were essential to prevent the number of attached electrons from fluctuating. Here, the point is the charging effect that blocks new electrons from adding onto materials of small size.

The charging effect is a purely classical electrostatic phenomenon. The essence is easily understood by considering the electrostatic charging energy of a metallic sphere of radius r. The capacitance of the sphere is readily calculated to be $C = 4\pi\varepsilon_0 r$. Here, the dielectric constant of vacuum is ε_0. Charging of the sphere with N electrons takes an electrostatic energy $E_c(N) = 1/2C(eN)^2 = (eN)^2/8\pi\varepsilon_0 r$. Therefore, work $\Delta E_c = E_c(N+1) - E_c(N) = e^2(2N+1)^2/8\pi\varepsilon_0 r$ is necessary to increase the number of electrons from N to $N + 1$. The addition of a new electron is blocked by this charging energy ΔE_c.

The charging energy ΔE_c increases as $1/r$ with decreasing r. This means that the charging effect becomes strong when materials are decreased to extremely small sizes. For example, Table 2.3 lists some numerically calculated ΔE_c of a metallic sphere for several values of r. Converting ΔE_c to temperatures (i.e., $\Delta E_c = kT$), where T is as high as 800 K for $r = 10$ nm. The electron cannot overcome this energy barrier at room temperature ($T = 300$ K), which keeps the number of electrons attached on the 10 nm size sphere constant (Figures 2.30 and 2.31).

The charging energy increases dramatically for nanosized materials as shown in Table 2.3. It yields the characteristic current–voltage (I–V) curves for a system with a nanosized particle between two electrodes. For example, let us consider a system in which a metal nanosphere is located between two electrodes. On both sides of the electrodes, the nanosphere is isolated from an electrode by a very thin insulating layer. In this system, no current flows when a small bias voltage is applied between the electrodes because the probability of electrons tunneling through the insulating layer is very low. Further, the bias voltage is too small for electrons to overcome the charging energy in order for the addition of an electron onto the sphere (i.e., $e|V| < e^2/2C$). Therefore, a zero-conductance plateau appears in the

Table 2.3 Numerically Calculated Charging Energy of a Metallic Nanosphere

R	$E_c = e^2/8\pi\varepsilon_0 r$ (eV)	T (K) ($E_c = kT$)
1 nm	0.72 eV = 720 meV	8300
10 nm	0.072 eV = 72 meV	830
100 nm	0.0072 eV = 7.2 meV	83
1 μm	0.72 meV	8.3
10 μm	0.072 meV	0.83
100 μm	0.0072 meV	0.083
1 mm	0.00072 meV	0.0083

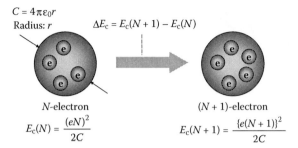

Figure 2.30 Charging effect of a nanosphere.

I–V characteristics of the system at a bias voltage $|V| < |V_{th}| = e/2C$, where C is the capacitance of the nanosphere. This zero-conductance plateau is called *Coulomb blockade*. However, the addition of an electron from an electrode to the nanosphere is allowed when the bias voltage exceeds a critical value ($|V| > |V_C|$). The electron added to the nanosphere tunnels through insulating barrier to the electrode on the other side. In this way, current starts to flow at $|V| > |V_C|$. As a result, a stepwise change appears at a bias voltage $|V| = |V_{th}|$ in the *I–V* characteristics.

An example is shown in Figure 2.32 [8]. In this study, gold nanoparticles of several nanometers in size are distributed on an insulating $CH_3(CH_2)_9SH(C_{10})$ monolayer formed on a Au(111) substrate as shown in the panel on the right-hand side. An STM tip is placed on top of each gold nanoparticle. The vacuum gap between the tip and the particle is about 1 nm. Here, the Au(111) substrate and the STM tip serve as electrodes, whereas

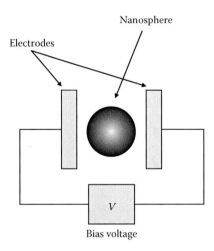

Figure 2.31 Experimental setup to observe the Coulomb blockade.

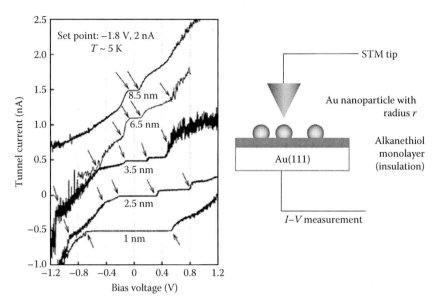

Figure 2.32 Coulomb blockade for Au nanoparticles in the tunneling gap of STM. (Reprinted from B. Wang et al., *Appl. Phys. Lett.* 77, 2000, 1179. With permission.)

the vacuum gap and the alkanethiol monolayer serve as insulating layers. The *I–V* characteristics are taken by applying a bias voltage between the STM tip and the substrate, and measuring the tunneling current at the STM tip at $T = 5$ K. In Figure 2.32, the Coulomb blockade plateaus (indicated by pairs of arrows) are clearly observed in the *I–V* characteristics. The length of the plateaus increases with decreasing size of the gold nanoparticles. For nanospheres, the capacitance *C* is proportional to $1/r$. Therefore, this is attributed to the increase in the charging energy with reduction in the size. Coulomb blockade is reported as is clearly observed at $T = 5$ K, but becomes obscure at higher temperatures (not shown in the figure). This is because the charging energy becomes comparable to the thermal fluctuation kT at higher temperatures.

When more than two nanospheres are bound between two electrodes, many plateaus appear and produce staircase-like *I–V* characteristics. This *Coulomb staircase* is also due to the charging effect of nanospheres.

2.6 Tunneling phenomena

If two materials spaced well apart are placed in an insulating environment, such as a vacuum, then they are electronically isolated. No current flows between the materials even if a bias voltage is applied between them. However, the situation changes dramatically if the distance between the materials is reduced to the nano- or subnanometer scale, in which case a current flows due to tunneling phenomena.

As an example, let us consider the tunneling current between two metallic materials located in a vacuum. For simplicity, we treat this phenomenon using a 1D model, as illustrated in Figure 2.33, which shows the band alignment of the two materials. In the figure, the gray area indicates valence electrons that occupy the energy states of the valence band up to the Fermi level (μ).

Usually, these electrons are confined in each material by the work function (ϕ). Since the work function and the occupied band width (i.e., the Fermi energy) are intrinsic properties of the materials, we put suffixes L and R on ϕ and μ, respectively, to distinguish the properties of the materials on the left and right sides in the figure.

The top-left panel in the figure indicates the situation where both materials on the left and right sides are isolated in a vacuum. In this case, the vacuum level aligns in the energy diagram because an electric field is not applied between the two materials. If we connect the two materials electrically by a wire, the energy diagram changes as shown in the top right panel of the figure. Here, the Fermi levels of the materials on the left and right align. Otherwise, the current flows between the materials in

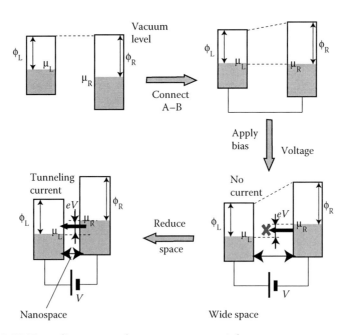

Figure 2.33 Tunneling current between two materials.

spite of the fact that the materials are only connected, with any external bias voltage being not applied. The alignment of the Fermi levels induces a gradient in the vacuum level, which hinders current flow between the two materials. This gradient represents the difference between the work functions of the left and right materials.

An application of an external positive/negative bias voltage V between the two materials stabilizes (or destabilizes) the chemical potential of the electrons in the material of one side more than that of the material of the other side. This situation is expressed by the downward (or upward) shift of the Fermi level of the material of one side by eV with respect to the Fermi level of the material in the other side, as illustrated in the bottom right panel in Figure 2.33. As a result, an energy region appears between the two Fermi levels in which the electron states are empty on one side, whereas the states are occupied on the other side. This enables the electrons to flow from the occupied states on one side to the empty states on the other side through the potential barrier between A and B. However, if the two materials are separated by a long distance, then the wave functions of the electrons decay and die out during the movement. Therefore, no current flows between the two separated materials, even if a bias voltage is applied between them. But, the movement of electrons becomes possible via electron tunneling if the distance between the two materials is decreased to the nanometer scale as illustrated in the bottom-left panel in Figure 2.33. Although the electron wave function decays in the gap between the two materials, a considerable part of the wave function still survives at the other side of the gap if the gap is narrow enough. Namely, a current can flow between the two materials with a very narrow gap due to electron tunneling phenomena.

Now, we will calculate an approximate value for the magnitude of the tunneling current. In the calculation, we assume that the two materials are close enough and that the material on the left side is positively biased, as illustrated in the bottom-left panel in Figure 2.33, where the bias voltage is V. In drawing the energy diagram in Figure 2.33, we assume that the electron DOS is constant (i.e., energy-independent) and express the electron DOS as the gray boxes. However, to make the calculation more realistic, we include the electron DOS of the materials on both sides as $\rho_R(E)$ and $\rho_L(E)$.

Under these conditions, the current flows by electron movement from the occupied states on one side to empty states on the other side via electron tunneling. The vacuum area between the materials acts as a potential wall. Referring to the Fermi level of the left-side material, the electrons feel the vacuum as a potential wall with a height of ϕ_L at the left side as illustrated in the top-right panel in Figure 2.34. The potential height of the vacuum region increases from left to right, and at the right side it becomes $\phi_R + eV$, as illustrated in the panel. The electrons tunnel through this potential barrier at the vacuum gap. Now, we define $T(E)$ as the tunneling probability of electrons of energy E. Under these assumptions, defining

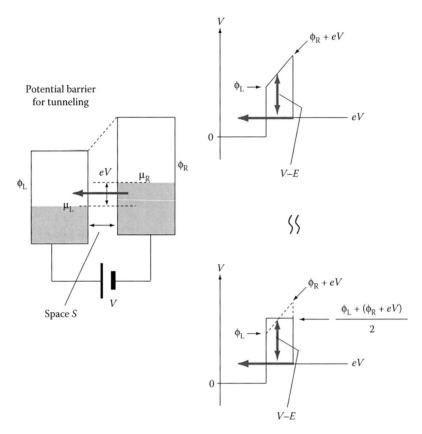

Figure 2.34 The energy diagram for the calculation of the tunneling current.

$f_R(E)$ and $f_L(E)$ as the Fermi distribution functions of A and B, the current I between the two materials is given by

$$
\begin{aligned}
I &= \frac{2e}{\hbar} \int_0^\infty f_L(E)\rho_L(E)\{1 - f_R(E)\}\rho_R(E)T(E)\,dE \\
&\quad - \int_0^\infty f_R(E)\rho_R(E)\{1 - f_L(E)\}\rho_L(E)T(E)\,dE \\
&= \frac{2e}{\hbar} \int_0^\infty \rho_L(E)\rho_R(E)\{f_L(E) - f_R(E)\}T(E)\,dE \\
&= \frac{2e}{\hbar} \int_{\mu_L}^{\mu_R} \rho_L(E)\rho_R(E)T(E)\,dE.
\end{aligned}
\tag{2.35}
$$

The first term describes the current flow from the occupied states on the left side to the empty states on the right side. The factor $f_L(E)\rho_L(E)$ describes the actual density of the occupied states on the right side, whereas $\{1 - f_R(E)\}\rho_R(E)$ describes the actual density of the empty states on the right side. Furthermore, the second term represents the current flow from the occupied states on the right side to the empty states on the left side. The net current is given by the difference of these two terms as expressed in the first line of Equation 2.35. The last line in Equation 2.34 is deduced by substituting the stepwise Fermi distribution function at $T = 0$ K [i.e., $f(E) = \theta(\mu - E)$] to the equation of the middle line. This is the final expression we wanted to deduce.

The tunneling current depends on the distance s between the two materials. The distance dependence is introduced by the tunneling probability $T(E)$ in the last expression in Equation 2.35. The distance dependence of $T(E)$ can be roughly estimated as follows: under a potential barrier $V > E$ (E is the energy of tunneling electrons), the wave function decays as $e^{-\kappa x}$ in the barrier region (equal to the space between neighboring materials). Here, x is the distance traveled and $\kappa = \sqrt{2m(V - E)}/\hbar$. As a result, the electron density $|\psi(x = 0)|^2$ on one side decays by a factor $e^{-2\kappa s}$ on the other side of the potential barrier $x = s$. In our system, the barrier height V changes as a function of the distance x in the vacuum tunneling gap as illustrated in the upper-right panel in Figure 2.35. This makes the estimation of the distance dependence of $T(E)$ difficult. However, in a crude approximation, the potential barrier $V(x)$ in the upper-right panel in Figure 2.35 can be replaced

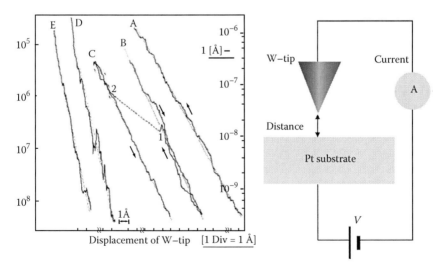

Figure 2.35 Experimentally observed distance dependence of the tunneling current. (Reprinted from G. Binnig et al., *Appl. Phys. Lett.* 40, 1982, 178. With permission.)

by the uniform barrier with an average height $V_{av} = \phi_R + eV + \phi_L/2$, as illustrated in the lower-right panel in Figure 2.34. For the uniform barrier of height V_{av}, the electron tunneling probability $T(E)$ is easily calculated to be $T(E) = |\psi(x = s)/\psi(x = 0)|^2 = \exp(-2\kappa s)$. More specifically,

$$T(E) = \exp\left\{-2\frac{\sqrt{2m(V_{av} - E)}}{\hbar} s\right\}. \qquad (2.36)$$

This expression for $T(E)$ is a good approximation of more rigorous results obtained using the WKB (Wentzel–Kramers–Brillouin) approximation.

Equation 2.36 shows that the tunneling probability decays exponentially with the distance s. This distance dependence is readily observed experimentally, and a few examples of tunneling current versus distance are shown in Figure 2.35 [9]. In the experiment, an STM tip was put very close to a Pt substrate surface and the tunneling current between the tip–surface was measured as a function of the tip–surface distance at a constant bias voltage. Although the condition of the tip produces scatter in the gradient, all the *I–s* curves reveal an exponential decay of the tunneling current with s, as predicted by Equation 2.36. It is noteworthy that the magnitude of the tunneling current is very sensitive to extremely small changes of the distance s on a subnanometer (Angström) scale. The sensitivity of $T(E)$ to s is also found numerically using Equation 2.36. Using this equation, T is calculated to be 0.38, 4.5×10^{-5}, 3.7×10^{-44} for $s = 0.1$, 1, and 10 nm for $V_{av} - E$ of 1 eV (a typical value in practical cases) in Equation 2.36. Although this equation includes crude approximations, the results demonstrate that the tunneling effect becomes dominant only at nano- and subnanometer scales.

AN ASIDE: STM

The STM is an example of technology that exploits tunneling phenomena. The STM is an instrument for characterizing the structure and electronic states at materials' surfaces. The experimental setup of a typical STM is shown in Figure 2.36. In an STM, an atomically sharp tip is placed in extremely close proximity to the surface of a conducting material. A bias voltage is applied between the tip and the surface, and the tunneling current across the vacuum gap is measured. Here, the tunneling current is very sensitive to the tip–surface distance as shown in the previous section. Therefore, the tip is fixed onto a piezoelectric actuator, which enables controlling its position with sub-Angström resolution in a direction perpendicular (z-direction) to the substrate. In addition, the tip is designed to move parallel to the

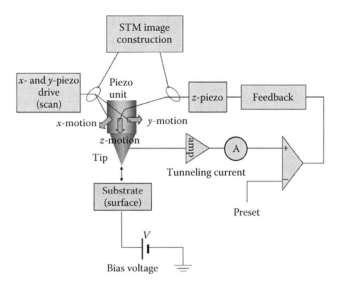

Figure 2.36 STM apparatus.

surface plane (x–y-plane) with sub-Angström resolution with x- and y-motion piezoelectric actuators. In STM, the surface is scanned by the tip using x- and y-motion piezoelectric actuators and the tunneling current is monitored at predefined coordinates as the tip is scanned across a sample surface. During the scan, the tunneling current changes due to changes in the tip–surface distance due to undulation in the surface. The changes in the tunneling current are quickly fed back to circuits applying voltages to the z-piezoactuator so as to keep the tunneling current constant by adjusting the tip–surface distance during the scan. Since the tunneling current is very sensitive to the tip–sample distance, the applied voltage to the z-piezo reflects any minute undulations such as the atom arrangements, and the steps and terraces of structures on surfaces. Changes in the applied voltage to the z-piezo are recorded at each tip position during the scan. Thus in STM, surface undulations are visualized as variations of the applied bias voltage to the z-piezo. Typically, the bias voltage, tunneling current, and the tip–surface distance are set to be 10 mV–2 V, 10 pA–1 nA, and 1 nm. The system is put onto a vibration isolator to cut mechanical noises and for the imaging at true atomic resolution.

STM is usually known as a tool to monitor surface topography. However, strictly speaking, the STM does not directly monitor topography, but the spatial distribution of electron DOS at surfaces. This is easily understood by using the simple 1D model for the tunneling phenomena discussed in the previous subsection. The model is schematically reproduced in Figure 2.37. Here, the right and left

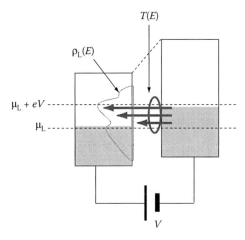

Figure 2.37 The energy diagram for the tunneling current in STM.

materials represent the electron bands of the tip and the sample in the STM, respectively. In the figure, a positive bias voltage V is applied to the sample on the left side with respect to the tip on the right side. As a result, the tunneling current I flows from the tip to the sample. For simplicity, we neglect the electron DOS $\rho_R(E)$ of the tip, and assume it is constant (i.e., E-independent). In this assumption, the tunneling current in Equation 2.35 is simplified as $I \propto \int_{\mu_L}^{\mu_L+eV}$ $\rho_L(E) \, T(E) \, dE$. This equation means that in STM only the electron DOS of the sample surface in the energy range $\mu_L \leq E \leq \mu_L + eV$ contributes to the tunneling current. Furthermore, the equation tells us that the contribution from each state is weighted by $T(E)$ in the tunneling current. On the other hand, $T(E)$ increases rapidly with E as predicted by Equation 2.36. Thus, to a rough approximation, the tunneling current in STM mainly reflects the surface electron DOS at $E = \mu_L + eV$. In the sample, the electrons occupy states where $E \leq \mu_L$; therefore, states where $\mu_L < E$ are empty. Thus, STM images taken at a positive bias voltage reflect the spatial distribution of empty state electrons at $E = eV$ with respect to the Fermi level at the surface. Also, STM images with negative sample bias voltages represent the spatial distribution of occupied state electrons at $E = -eV$ with respect to the Fermi level at the surface.

For example, an STM image of the submonolayer of Ag-covered Si(111) surface is shown in Figure 2.38. The unit cell of this surface atomic arrangement has a size of 3×1 with respect to that in bulk Si crystals. Therefore, this surface is called Si(111)3×1-Ag. The figure shows the dual-bias STM image, in which the sample bias voltage is

Si(111)3 × 1-Ag

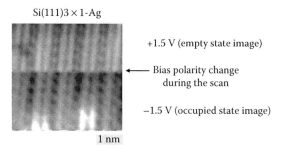

+1.5 V (empty state image)

← Bias polarity change during the scan

−1.5 V (occupied state image)

1 nm

Figure 2.38 Dual-bias STM image of the Si(111)3 × 1-Ag surface.

switched from +1.5 to –1.5 V during the scan. As a result, the upper-half of the image represents the spatial distribution of empty states, whereas the lower-half represents that of the occupied state at the surface. The size of the image is 5 × 5 nm. In both halves, atomically resolved patterns are observed. However, the patterns are staggered at the line where the bias voltage polarity is changed, although the surface atomic arrangement is the same in both halves. This is due to differences of the spatial distribution of the empty and occupied electronic states associated with the same surface atomic arrangement. As shown in this example, an STM represents the spatial distribution of electronic states at surfaces. It reveals the atomic arrangement if electron states eventually have the same spatial distribution as surface atoms.

2.7   Limiting factors for size effects

2.7.1   Thermal fluctuation

Many remarkable electron features originate from quantum confinement. Notably, quantum phenomena are usually studied at very low temperatures because the quantum nature is not observed by thermal fluctuation at elevated temperatures. In previous sections, we did not explicitly consider temperature under which the nanomaterials are placed. Namely, we treated the phenomena at $T = 0$ K. Here, we consider the effects of temperature on the quantization of electron states of nanomaterials. As we will find in this subsection, the smaller the size, the more robust the nanomaterials become against thermal fluctuations.

An electron—the main player in quantum confinement—is a fermion. Statistical physics tells us that the probability $f(E)$ of a fermion occupying an electronic state with energy E is given by

$$f(E) = \frac{1}{\exp\{(E-\mu)/kT\}+1} \tag{2.37}$$

at temperature T. Here, μ is the chemical potential (the Fermi level) of the system. The function $f(E)$ is called the Fermi distribution function. Thermal fluctuations originate from the T dependence of the Fermi distribution function.

The Fermi distribution function $f(E)$ is depicted as a function of energy E for several temperatures T in the left panel of Figure 2.39. As shown in the figure, $f(E)$ is a step function with a sharp knee at $E = \mu$ at $T = 0$ K. In this case, there are no fluctuations in the occupation of the energy states. Consequently, the occupation probability of states at $E \leq \mu$ is unity, while that of states at $E > \mu$ is zero. However, the rapid stepwise change of $f(E)$ at $E = \mu$ is smeared out with increasing temperature T, as shown in the figure. At temperature T, the transition of $f(E)$ from unity to zero occurs gradually within the energy range $E \approx \mu \pm kT$. This thermal broadening makes it difficult to distinguish between the occupied and unoccupied quantized states in the energy range $|E - \mu| < kT$ as illustrated in the right panel in Figure 2.39. For example, let us assume that $n = 1$ and $n = 2$ QWSs are below and above the Fermi level, respectively. Here, if the energy spacing between $n = 1$ and $n = 2$ QWSs is smaller than the thermal broadening width kT (the most right-side case in the right panel), it becomes meaningless to recognize the $n = 1$ QWS as a fully occupied state and $n = 2$ as a fully empty state. Meanwhile, the origin of properties such as quantized conductivity is based on counting the number of effective quantized channels. Thermal broadening makes the distinct property of quantized states ambiguous. In this manner, thermal broadening hinders the manifestation of quantum properties.

The magnitude of this effect depends on the thermal broadening width kT relative to the energy difference (ΔE) between the quantized

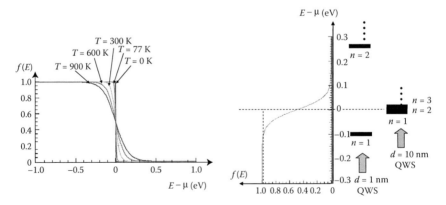

Figure 2.39 The Fermi distribution function. Left: T-dependent change of $f(E)$. Right: a comparison of the transient region of $f(E)$ and the energy spacing between QWSs.

states around the Fermi level as shown in the right panel in Figure 2.39. Thermal fluctuations are negligible for $\Delta E \gg kT$, but extremely pronounced for $\Delta E \approx kT$. Let us compare kT with ΔE of a simple 1D infinite potential well of width d. In the well, the electrons are quantized into discrete energy levels as

$$E_n = \frac{\hbar^2}{2m}\left(\frac{n\pi}{d}\right)^2. \tag{2.38}$$

Thus, the energy spacing between quantized states is of the order of $\Delta \approx \hbar^2/2m(\pi/d)^2$. It is calculated to be 370 meV for $d = 1$ nm, and 3.7 meV for $d = 10$ nm, as illustrated in the right panel in the figure. On the other hand, the thermal fluctuation kT is 26 meV at $T = 300$ K, 6.6 meV at $T = 77$ K, and 0.34 meV at $T = 4$ K. Therefore, $\Delta E \gg kT$ holds for a quantum well of $d < 1$ nm even at $T = 300$ K.

The above calculation demonstrates that the size of the material is the key to overcome thermal broadening. In conventional macroscopic materials, it is impossible to make their size to be in the nanometer region. In this case, the energy spacing between the QWSs is very small and low temperatures are necessary to manifest their quantum properties. However, materials with dimensions on the nanometer scale have a chance to overcome thermal broadening. Especially, materials of true nanometer scale are promising to realize their quantized functions even at room temperature.

2.7.2   Lifetime broadening effect

The energy–time uncertainty principle $\Delta E \cdot \Delta t \geq \hbar/2$ holds between an energy E and time t. This means that δ-function-like sharp peaks in the energy spectrum (i.e., $\Delta E = 0$) are guaranteed only in cases of $\Delta t = \infty$. However, if the lifetime of an electron in a quantized state is finite (i.e., $\Delta t \neq \infty$), then quantized peaks become broader (i.e., $\Delta E > 0$). The quantum nature of nanomaterials originates due to the sharpness of quantized states. Thus, this lifetime broadening hinders the manifestation of the quantum nature of nanomaterials.

In the quantum limit ($\Delta E \cdot \Delta t \geq \hbar/2$), ΔE is of the order of 10^{-10}, 10^{-7}, 10^{-4}, and 10^{-1} eV for $\Delta t = 1 \times 10^{-6}$ s (1 µs), 1×10^{-9} s (1 ns), 1×10^{-12} s (1 ps), and 1×10^{-15} s (1 fs), respectively. Meanwhile, the energy spacing between quantized states is of the order of 100 meV in a 1D infinite potential well of width 1 nm. In this example, the lifetime Δt must be longer than 1 fs to prevent two adjacent quantized peaks from overlapping due to lifetime broadening.

The lifetime of electrons is limited mainly by inelastic scattering with impurities, phonons (vibrating lattice atoms), and other electrons in materials. Electrons in quantized states escape gradually by inelastic scattering processes. Electrons also escape from quantized states when the confinement potential is too low. This leakage of electrons shortens the lifetime of quantized states and makes their δ-function-like peaks broader in energy spectra.

For example, the effect of electron leakage on the broadening of quantized states is examined by considering electrons confined in a 1D quantum well with leaky boundaries (the upper panel in Figure 2.40). The electrons make round trips in the well by repeatedly bouncing from boundaries.

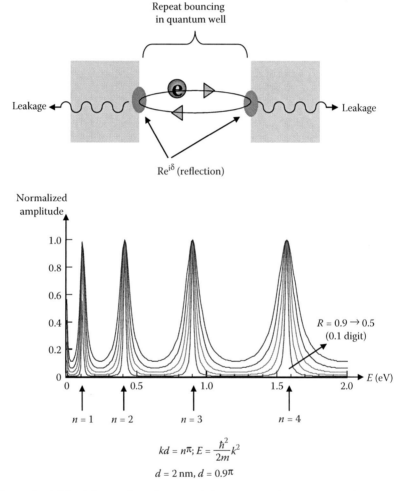

$$kd = n\pi;\ E = \frac{\hbar^2}{2m}k^2$$

$$d = 2\,\text{nm},\ d = 0.9\pi$$

Figure 2.40 The lifetime broadening of the quantized states in a 1D quantum well with leaky boundaries.

However, electrons are gradually lost from the quantum well at every bounce through the leaky boundaries. As a result, electrons possess a finite lifetime in quantum wells. Here, let us assume the width of a quantum well to be d, the wavevector of electrons k, the reflectance of electrons at a boundary R, and the phase change on bouncing δ. As we discussed in Section 2.3.5, the electron gains a phase $(2kd + 2\delta)$ during one round trip. Therefore, the electron wave function $\exp(ikx)$ (at the position x) evolves to be $R^2 \exp\{ikx + (2kd + 2\delta)\}$ after one round trip. Here, R^2 comes from bouncing at the leaky right and left boundaries during one round trip. Also, the wave function evolves as $R^4 \exp\{ikx + (4kd + 4\delta)\}$ after two round trips. All these waves interfere at the position x in the quantum well. As a result, the electron wave function is totally described as follows:

$$\psi(x) = \exp(ikx)\{1 + R^2 \exp i(2kd + 2\delta) + R^4 \exp i(4kd + 4\delta) + \cdots\}$$

$$= \sum_{n=0}^{\infty} R^{2n} \exp i\{kx + n(2kd + 2\delta)\} = \frac{1}{1 - R^2 \exp i(2kd + 2\delta)} \exp(ikx). \quad (2.39)$$

Therefore, the amplitude of the electron density $|\psi(x)|^2$ is given by $|1/(1 - R^2 \exp i(2kd + 2\delta))|^2$. The amplitude is a function of the wavevector k. However, the wavevector k is related to the energy of an electron E by the dispersion relationship of freely moving electrons given by $E(k) = \hbar^2/2mk^2$. Thus, the amplitude can be regarded as being a function of the electron energy E. Based on these arguments, the amplitude is shown as a function of E in the lower panel in Figure 2.40.

The amplitude has several peaks corresponding to QWS. Actually, if the confinement potential is infinite and allows no leakage (namely, $R = 1$ and $\delta = \pi$), then the amplitude becomes a δ-function that has sharp peaks at $kd = l\pi$ ($l = 1, 2, 3, \ldots$). The condition $kd = \pi, 2\pi, 3\pi, \ldots$ corresponds to $d = \lambda/2\pi \times \pi, 2\pi, 3\pi, \ldots = \lambda/2, 2\lambda/2, 3\lambda/2, \ldots$. Namely, $l = 1, 2, 3, \ldots$ gives the first, second, third, ... QWSs obtained by solving Schrödinger's equation for infinite potential barrier boundary conditions.

However, $R < 1$ and $\delta \neq \pi$ for the leaky boundaries. The normalized amplitude is plotted as a function of energy for $R = 0.9, 0.8, 0.7, 0.6,$ and 0.5 in the lower panel of Figure 2.40. As shown in the figure, the larger the leakage [$\propto (1 - R)$], the broader is the peaks of the QWS.

2.8   Electronically induced stable nanostructures

In the previous subsections, we considered the characteristic electron states and properties of nanomaterials. The electron states depend strongly on the size and shape of nanomaterials. Conversely, a subtle change of the size and shape modifies the electron states drastically in nanomaterials. In some cases, the electron energy has a local minimum for specific (magic) sizes and shapes of nanomaterials. Here the nanomaterials of the magic size and shape are stable and robust against fluctuations in size and shape. This electronically induced stability can be utilized to make the size and shape of nanomaterials uniform. In this subsection, we give two examples of electronically induced stable nanostructures: the magic number in clusters and magic thicknesses in nanofilms.

2.8.1   Magic numbers in clusters

A cluster is an ensemble composed of approximately tens to several hundreds of bound atoms. The number of atoms in clusters is larger than that in molecules, but is much smaller than that (more than million atoms) in bulk solids. The size of small clusters is typically in the nanometer region (nanoclusters). The size and shape of nanoclusters are governed by two factors: the surface energy and the closed-shell structure.

As we learned in Section 2.4, most of the atoms are located on surfaces in materials of nanometer scale (Figure 2.26). In addition, the surfaces are energetically unfavorable. Thus, nanoclusters generally tend to minimize their surface area. A sphere has the smallest surface area among solids. In this respect, nanoclusters tend to be spherical if the surface energy is isotropic.

However, the surface energy of crystalline materials is anisotropic. The surface energy changes depending on the surface orientation, and nanoclusters are surrounded by several (low-indexed) low-energy surfaces and take polygonal shapes. Usually, nanoclusters are icosahedral in shape because the icosahedral is a polygon that has a shape closest to that of a sphere as illustrated in the left panel in Figure 2.41 [10]. Meanwhile, the icosahedral clusters can be constructed only by a specific number of atoms, such as 13, 55, 92, ... , as illustrated in the right panel of Figure 2.41. By this reason, nanoclusters have a tendency to favor icosahedral shapes with such specific (magic) numbers of atoms as 13, 55, 92, ... In these icosahdral clusters, the ratio of surface/bulk atoms is 12/1, 42/13, 92/55, ... The icosahedral shape minimizes the area of the surfaces at which most of the atoms exist.

Apart from surface energy minimization, nanoclusters also favor electronically stable closed-shell structures. Since nanoclusters are a 0D system, they have an atomic-like core–shell electronic structure with principal, azimuthal, and magnetic quantum numbers. An electron shell consists of orbitals associated with the same principal quantum number, where the K shell is formed of 1s orbitals (the principal quantum number = 1). The 2s and 2p orbitals (the principal quantum number = 2) make up the L shell, and so on. Now, the chemical reactivity is mainly determined by the occupation of electronic states in the outermost shells. Atoms whose outermost electron shell is fully occupied are called rare gases (e.g., Ar, Kr, etc.) and are energetically stable. This situation is known

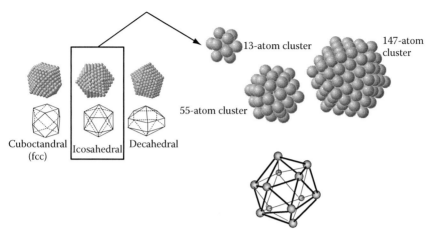

13-atom cluster

147-atom cluster

55-atom cluster

Cuboctandral (fcc)   Icosahedral   Decahedral

Figure 2.41 Icosahedral nanoclusters. (Adapted from http://phys.canterbury. ac.nz/research/solid_state/clusters.html. With permission.)

as a closed-shell structure. Just as atoms, nanoclusters with closed shells are stable. On the other hand, the number of electrons that occupy the electron states in their shells is determined by the number of atoms forming the clusters. Thus the nanoclusters tend to include specific (i.e., magic) number of atoms, which produce closed-shell electron structure.

As an example of magic numbers in nanoclusters, theoretical and experimental results for Na clusters are shown in Figure 2.42. The theoretically calculated shell structure [11] is listed in the left panel. The energy levels of the shells increase from bottom to top. The figure beside each shell shows the total number of electrons necessary to make the corresponding shell structure closed. The right panel represents the experimentally observed appearance frequency of Na clusters including N atoms [12]. Remarkably, sharp peaks are observed at $N = 8$, 20, and 40. This means that Na clusters with these magic numbers are very stable. In Na clusters, each Na atom supplies one valence electron. Therefore, the number of atoms N is equal to the number of electrons in the cluster. In this respect, the experimentally observed magic numbers in Na clusters can be compared with the theoretically calculated number of electrons required to make the closed-shell electronic structure. Comparing the theoretical and experimental results, we notice that the 1p, 2s, and 2p shells are closed at magic numbers of $N = 8$, 20, and 40, respectively. This demonstrates that the stability of the magic size clusters originates from the electronically closed-shell structure.

Both the preference of the icosahedral shape and the electronic closed-shell structure act to make magic clusters stable in some cases.

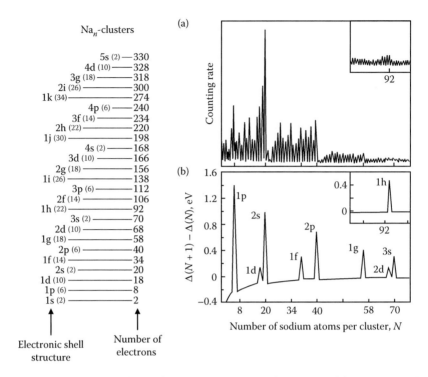

Na$_n$-clusters

(a)

Electronic shell structure	Number of electrons

5s (2) — 330
4d (10) — 328
3g (18) — 318
2i (26) — 300
1k (34) — 274
4p (6) — 240
3f (14) — 234
2h (22) — 220
1j (30) — 198
4s (2) — 168
3d (10) — 166
2g (18) — 156
1i (26) — 138
3p (6) — 112
2f (14) — 106
1h (22) — 92
3s (2) — 70
2d (10) — 68
1g (18) — 58
2p (6) — 40
1f (14) — 34
2s (2) — 20
1d (10) — 18
1p (6) — 8
1s (2) — 2

Figure 2.42 Magic numbers of Na nanoclusters. (Left: Reprinted from X. Li et al., *Phys. Rev. Lett.* 81, 1998, 1909. With permission; Right: Reprinted from W.D. Knight et al., *Phys. Rev. Lett.* 52, 1984, 2141. With permission.)

A good example is the extreme stability of negatively ionized Al_{13}^- against oxidation. The experimentally observed appearance frequency of Al_n^- clusters is shown in the left panel of Figure 2.43 [13]. Before oxidation, many peaks are observed in the top graph. However, only several peaks survived after oxidation (the middle and bottom graphs). In particular, the peak at $n = 13$ becomes remarkably sharp after hard oxidation (the bottom graph). This shows the extreme stability of Al_{13}^- against oxidation. The theoretically calculated electronic shell structure is shown in the right panel in Figure 2.43 [14]. The number of electrons to make the shell closed and the corresponding cluster structure are indicated beside the shells. Since one Al atom has three valence electrons, the negatively charged 13 Al atom clusters have $3 \times 13 + 1 = 40$ valence electrons. The right panel in the figure indicates that the 40 electrons complete the 2p closed-shell structure. In addition, the 13 Al atoms are able to construct the icosahedral shape as illustrated in the figure. The Al_{13}^- cluster is stable because of its icosahedral shape and due to the electronic closed-shell structure.

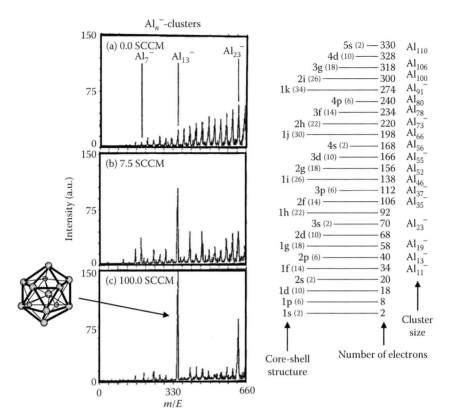

Figure 2.43 The stability of Al_{13}^- magic cluster against oxidation. (Right: Reprinted from X. Li et al., *Phys. Rev. Lett.* 81, 1998, 1909. With permission.)

2.8.2   Electronic growth

The second example of the electronically induced stable nanostructures is the so-called electronic growth mechanism for atomically flat, ultra-thin metal films with uniform thicknesses. The growth of a metal film is usually governed by thermally activated kinetic processes such as adsorption, diffusion, nucleation, and so on, at high temperatures. All these processes are stochastic and proceed locally. Therefore, it is impossible to realize thickness uniformities for thin metal films grown on substrates. However, atomically flat ultrathin metal films with uniform thicknesses have recently been produced by a two-step (a deposition at a low temperature and subsequent slow annealing to room temperature) growth procedure. Here, the thickness uniformity is explained in terms of a theory based on the thickness-dependent changes of the electronic energy of thin films. The the rmally activated kinetic processes are supposed to be suppressed in the two-step growth procedure, and electronic factors play a dominant role during film growth.

Before discussing the mechanism, we show an example of the surface morphology of an ultrathin metal film grown by a two-step process. Figure 2.44 shows the STM images of the morphology of Ag films grown by conventional room temperature deposition (the upper panel) and by the two-step growth method (the middle panel; the deposition temperature was 80 K) [15]. The image size is 300×300 nm^2. Room temperature growth produces films that are torn into pieces by deep grooves for all thicknesses.

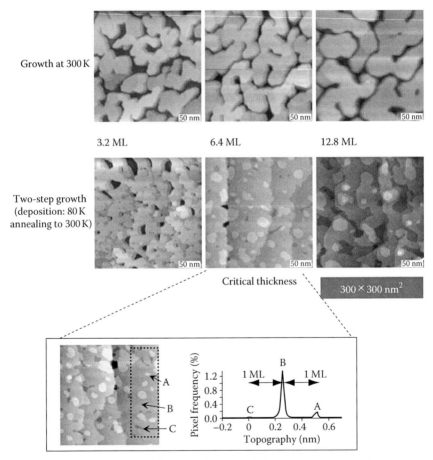

Figure 2.44 STM images of the morphology of Ag films grown by the conventional room temperature deposition (upper panel) and by the two-step growth method (middle panel). The lower panel shows the frequency height of the 6.4 ML thick Ag film grown by the two-step method. (Reprinted from M. Miyazaki and H. Hirayama, *Surf. Sci.* 602, 2008, 276. With permission.)

Each piece is flat, but the height differs from piece to piece on the atomic scale. In contrast, the two-step grown films do not have grooves for coverage above ~5 ML. The thickness uniformity is greatly improved by the two-step growth process. In particular, the 6.4 ML thick Ag films have nearly perfect thickness uniformity. The film had a unique height over almost all of the samples, as shown in the frequency height graph in the lower panel. This means that 6 ML is the magic thickness for Ag films grown by two-step processes, in which the preferable film thickness is determined by electronic factors.

The magic thickness of ultrathin metal films is explained by electronic growth theory [16]. In atomically flat ultrathin metal films, the valence electrons are confined in the film and are quantized along the surface direction as illustrated in Figure 2.7. However, a part of the confined electrons spill out from the metal films at the film/substrate interface because the confinement potential at the interface is finite. This spillage generates an interface electric dipole layer. The theory considers changes of these factors with film thickness.

Assuming a free-electron-like dispersion, the energies of quantized states are given by $E_n = \hbar^2/2m(n\pi/d)^2$, where $n = 1, 2, 3, \ldots$ and d the thickness as shown in Equation 2.19. Therefore, the energy of confined electrons E_{QWS} behaves as $E_{QWS} \propto 1/d^2$ with increases in the thickness d. Meanwhile, the spillage of confined electron increases with an increase in the electron confinement energy. The spillage-induced interface dipole layer reduces the electrostatic potential for electrons confined in the metal film. The film thickness dependence of the spillage-induced change of the electrostatic potential can be approximated as $E_C \propto -1/d$ [17]. Simply describing the total energy of the electrons in the film as $E_{el} = E_{QWS} + E_C = \alpha/d^2 - \beta/d$ (α, β: positive coefficients), one can expect that the energy is minimized at a specific (magic) thickness of $d_{critical} = 2\alpha/\beta$, as illustrated in Figure 2.45.

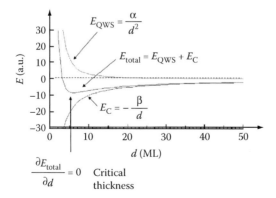

Figure 2.45 Thickness-dependent change of the energy of electrons in ultrathin metal films.

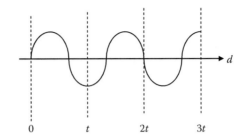

$$\frac{HOS + LUS}{2} - E_f$$

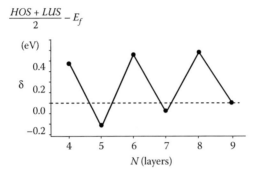

Figure **2.46** The magic thickness for ultrathin Pb(111) films. (Reprinted from W.B. Su et al., *Phys. Rev. Lett.* 86, 5116, 2001. With permission.)

A more detailed calculation of electronic growth theory predicts a magic thickness of 5 ML for Ag films, which is in satisfactory agreement with the experimental results. This demonstrates that the magic thickness of ultrathin metal films is induced electronically.

In addition to the factors discussed above, the so-called Friedel oscillation plays an important role in determining the magic thickness of ultrathin metal films in cases where the Fermi wavelength and metal layer thickness are commensurate. For example, the Fermi wavelength of a valence electron λ_F satisfies the relation $2t = 3 \times (\lambda_F/2)$ for a Pb(111) monolayer of thickness t. Because of this commensurate relation, only Pb films with even layer heights can accommodate the last node of the electron waves at the Fermi energy, as illustrated in the top panel of Figure 2.46. For this reason, the Pb film stability δ oscillates with a 2 ML period with increasing film thickness, as illustrated in the lower panel of Figure 2.46. Actually, the magic thickness is observed to appear with 2 ML period experimentally [18].

PROBLEMS

2.1 What is the equivalent of 1 electron volt (eV) in Joules (J)? Also, what is the equivalent of 1 eV in Kelvin (K)?

2.2 Describe the relationship between the lattice constant and de Broglie wavelength of normal metals.

2.3 Deduce the numerical relationship between the width d (in nm) and the energy separation (in eV) between the $n = 1$ and $n = 2$ quantized states in a 1D quantum well with the infinite potential barriers. For example, analyze several values of d in the range 1–100 nm.

2.4 Verify the dimensions of 1D, 2D, and 3D electron DOS (per energy).

2.5 Calculate the quantized conductance $2e^2/h$ in units of Ω^{-1}.

2.6 Calculate several of the lowest QWSs for a 2D square ($d = 10$ nm) and for a 2D cylindrical well ($a = 5$ nm). In both cases, compare the energy separation between the quantized states with the thermal energy, kT for $T = 300$ K.

2.7 Calculate several of the lowest QWSs for a 1D finite quantum well with $V_C = 5$ eV and $d = 1$ nm. Compare the results with those of a 1D infinite quantum well of the same size.

2.8 Calculate the tunneling probability for $V_{av} = 5$ eV and s in the range 0.1–10 nm. Show the s dependence in the form of a graph.

References

1. M. Miyazaki, Master thesis, Tokyo Institute of Technology, Tokyo, 2008.
2. I. Matsuda, T. Ohta, and H.W. Yeom, *Phys. Rev. B* 65, 2002, 085327.
3. G. Rubio, N. Agrait, and S. Vieira, *Phys. Rev. Lett.* 76, 1996, 2302.
4. J. Kliewer, R. Brendt, and S. Crampin, *New J. Phys.* 3, 2001, 22.
5. S. Hla, K. Braum, and K.H. Rieder, *Phys. Rev. B* 67, 2003, 201402.
6. N. Lang and W. Kohn, *Phys. Rev. B* 1, 1970, 455.
7. N.D. Lang, *Solid State Phys.* 28, 1973, 255.
8. B. Wang, X. Xiao, S. Huang, P. Sheng, and J. Hou, *Appl. Phys. Lett.* 77, 2000, 1179.
9. G. Binnig, H. Rohler, Ch. Gerber, and W. Weibel, *Appl. Phys. Lett.* 40, 1982, 178.
10. http://phys.canterbury.ac.nz/research/solid_state/clusters.html
11. X. Li, H. Wu, X. Wang, and L. Wang, *Phys. Rev. Lett.* 81, 1998, 1909.
12. W.D. Knight, K. Clemenger, W. deHeer, W. Saunders, Y. Chou, and M. Cohen, *Phys. Rev. Lett.* 52, 1984, 2141.
13. R. Leuchtner, A. Harms, and A. Castleman, *J. Chem. Phys.* 91, 1989, 2753.
14. X. Li, H. Wu, X-B. Wan, and L-S. Wan, *Phys. Rev. Lett.* 81, 1998, 1909.
15. M. Miyazaki and H. Hirayama, *Surf. Sci.* 602, 2008, 276.
16. Z. Zhang, Q. Niu, and C-K. Shih, *Phys. Rev. Lett.* 80, 1998, 5381.
17. H. Hirayama, *Surf. Sci.* 603, 2009, 1492.
18. W.B. Su, S.H. Chang, W.B. Jian, C.S. Chang, L.J. Chen, and T.T. Tsong, *Phys. Rev. Lett.* 86, 2001, 5116.

chapter three

Optical properties and interactions of nanoscale materials

In this chapter, we describe changes that occur in the optical properties of materials when their sizes are decreased to the nanometer scale. Intriguingly, such changes have been known and used for centuries. Some of the most dramatic changes in the optical properties of materials—perceived by the human eye as changes in color—occur on scales of 0.1–10 nm. In the first part of this chapter, we describe size-dependent optical properties from the perspective of absorption and emission (Section 3.1), where the size-dependent properties of materials are classified according to the types of excitons involved in interactions. Size effects in absorption and scattering are treated in Section 3.2 by introducing the results of the general theory of Mie. This is followed by a description of localized plasmons in metallic nanoparticles and their applications in surface-induced enhancement of Raman scattering. The next two sections focus on nanometer-scale interactions. The interactions are classified in terms of particle–particle phenomena (discussed in Section 3.3), which include mainly strongly distance-dependent energy transfer processes as well as photo-induced charge transport, and particle–light interactions. The latter (described in Section 3.4) is due to modifications of the properties of light in confined geometries such as optical microcavities and the presence of dielectric interfaces. Both these effects significantly affect light emitted by materials.

3.1 Size-dependent optical properties: Absorption and emission

Absorption and emission of light are the most important effects of the size-dependent optical properties of nanoscale materials. This section begins with an introduction to the basic concepts of the semiclassical description of matter–light interaction by derivation of the rate of optical transitions. Classical treatment of the interaction between an atom or molecule and light based on the Lorentz oscillator model correctly predicts the dispersion relationship for the real part of the refractive index, and by introducing an arbitrary damping of electron oscillation, it explains the frequency dependence of the imaginary part of the refractive index in the form of the Lorentzian absorption line shape. For an understanding of the physical meaning of absorption processes, it is necessary to consider a quantum-mechanical description of matter–light interaction. In the semiclassical approach, which is introduced here, molecules are treated quantum mechanically and the related equations of motion are based on the time-dependent Schrödinger equation. Electromagnetic radiation is treated classically based on Maxwell's equations. This approach is sufficient to explain the processes of light absorption and stimulated emission. For a full treatment of spontaneous emission, it is necessary to include a theoretical treatment of the quantization of electromagnetic fields.

A theoretical introduction is followed by a general description of excitons. The size-dependent properties of materials are due to quantum confinement of Wannier excitons in inorganic semiconductors on scales of 0.5–20 nm, due to electron delocalization in organic conjugated systems on scales of 0.5–5 nm, and due to coherent-length phenomena in Frenkel excitons on scales of 1–100 nm. Finally, effects arising from a limited number of optical electrons in individual nanoscale emitters are briefly summarized.

3.1.1   Basic quantum mechanics of linear optical transitions

In the quantum-mechanical description, the processes of light absorption and emission involve transitions between two discrete quantum states of the system (atoms or molecules). The states differ by the energy of a single valence electron that responds to the electromagnetic perturbation of light. The quantum-mechanical treatment of optical transitions is specific in the sense that the energies and wave functions of the two states are given beforehand, and we are solving the problem of transition rates between these two energy levels. The transition rates give the probabilities per second of the transition taking place, thus yielding the absorption strength and emission intensity.

We begin by describing the system in terms of a set of two time-dependent, one-electron wave functions $\Psi_1(r, t)$ and $\Psi_2(r, t)$ corresponding to the ground and excited electron states, in the form

$$\Psi_1(r,t) = \psi_1(r)\exp\left(-i\frac{E_1}{\hbar}t\right) \quad \text{and} \quad \Psi_2(r,t) = \psi_2(r)\exp\left(-i\frac{E_2}{\hbar}t\right), \quad (3.1)$$

where $\psi_1(r)$ and $\psi_2(r)$ are time-independent functions of the electron position r and E_1 and E_2 are the energies of the two states given by the energy eigenvalue equations with a unperturbed Hamiltonian operator H,

$$H\Psi_i(r,\ t) = E_i\Psi_i(r,\ t). \tag{3.2}$$

The energy difference corresponds to the transition frequency ω_0,

$$E_2 - E_1 = \hbar\omega_0. \tag{3.3}$$

The wave function describing the system as a whole can be written as a time-dependent linear combination of the wave functions of states 1 and 2,

$$\Psi(r,\ t) = C_1(t)\Psi_1(r,\ t) + C_2(t)\Psi_2(r,\ t). \tag{3.4}$$

The function must be normalized at all times, which leads to the following condition for the time-dependent coefficients $C_1(t)$ and $C_2(t)$:

$$\int \Psi^*(r,\ t)\Psi(r,\ t)\,dr = |C_1(t)|^2 + |C_2(t)|^2 = 1. \tag{3.5}$$

Interaction of the system with electromagnetic radiation results in a change of the molecular potential energy. This change can be described by introducing a new Hamiltonian operator $H + H'$, where H is the unperturbed part from Equation 3.2 and H' is the part describing the interaction. The corresponding Schrödinger equation (Chapter 1) can be written as

$$i\hbar\frac{\partial}{\partial t}(C_1(t)\Psi_1(r,t) + C_2(t)\Psi_2(r,t)) = (H + H')(C_1(t)\Psi_1(r,t)$$
$$+ C_2(t)\Psi_2(r,t)). \tag{3.6}$$

This equation can be simplified by using the known stationary solutions of the unperturbed states in the form

$$i\hbar\frac{\partial \Psi_i(r,t)}{\partial t} = H\Psi_i(r,t). \tag{3.7}$$

These terms cancel out on both sides of Equation 3.6 and result in the expression:

$$i\hbar\left(\Psi_1\frac{dC_1}{dt} + \Psi_2\frac{dC_2}{dt}\right) = H'(C_1\Psi_1 + C_2\Psi_2), \tag{3.8}$$

where the space and time-dependence notation have been omitted for simplicity. Multiplying Equation 3.8 from the left by Ψ_1^*, integrating over all space, and making use of the normalization and orthogonality conditions gives on the left-hand side:

$$i\hbar\left(\int \Psi_1^*\Psi_1\, dr\frac{dC_1}{dt} + \int \Psi_1^*\Psi_2\, dr\frac{dC_2}{dt}\right) = i\hbar\frac{dC_1}{dt}, \tag{3.9}$$

and on the right-hand side:

$$C_1\int \psi_1^*\exp\left(i\frac{E_1}{\hbar}t\right)H'\psi_1\exp\left(-i\frac{E_1}{\hbar}t\right)dr$$
$$+ C_2\int \psi_1^*\exp\left(i\frac{E_1}{\hbar}t\right)H'\psi_2\exp\left(-i\frac{E_2}{\hbar}t\right)dr. \tag{3.10}$$

Combining Equations 3.9 and 3.10 gives

$$i\hbar\frac{dC_1}{dt} = C_1\int \psi_1^*H'\psi_1\, dr + C_2\exp(-i\omega_0 t)\int \psi_1^*H'\psi_2\, dr, \tag{3.11}$$

and similarly,

$$i\hbar\frac{dC_2}{dt} = C_2\int \psi_2^*H'\psi_2\, dr + C_1\exp(i\omega_0 t)\int \psi_2^*H'\psi_1\, dr. \tag{3.12}$$

Equations 3.11 and 3.12 are time-dependent Schrödinger equations with coefficients C_1 and C_2, which in principle can be solved for any given form of the interaction Hamiltonian operator H'. Solutions of the equations give the time-dependent probabilities of finding the system in ground or excited electron states and thereby describe interactions of the system with light, which lead to optical transitions.

Before proceeding with solving Equations 3.11 and 3.12, we must specify the form of the interaction Hamiltonian operator H'. In classical physics, as a consequence of Maxwell's equations, the oscillating electric field constituting light is expressed as a harmonic function of position z and time t:

$$\mathbf{E} = \mathbf{E}_0\cos(kz - \omega t). \tag{3.13}$$

Here, ω is the angular frequency of light and k is the propagation number. The vector notation represents the direction of oscillation of the electric field, which is equal to the direction of light *polarization*. For optical frequencies, the size of typical molecules is much smaller than the wavelength, and spatial variations of electric fields across the molecule are negligible. We may therefore neglect the space dependence of the electric field and use $\mathbf{E} = \mathbf{E}_0\cos(\omega t)$.

The electric field interacts with all charges in the system, making the theoretical treatment very complicated. For a good approximation, it is sufficient to expand the potential energy of the charges in the electric field into a multipole expansion and retain only the electric dipole term for one electron. Considering a molecule, the dipole moment is $\mu = -er$, where e is the electron charge and r its position vector. Interaction with the electric field of light changes the corresponding potential energy by $\mathbf{E} \cdot \mu$. The interaction Hamiltonian operator is then written as

$$H' = \mathbf{E}_0 \cdot \mu \cos(\omega t) = -e\mathbf{E}_0 \cdot \mathbf{r} \cos(\omega t). \tag{3.14}$$

The operator H' has an odd parity $H'(-\mathbf{r}) = -H'(\mathbf{r})$, meaning that $\psi_i^* H' \psi_i$ must also have odd parity and $\int \psi_i^* H' \psi_i \, dr = 0$. The Schrödinger equations (Equations 3.11 and 3.12) for the coefficients C_1 and C_2 then simplify to

$$i\hbar \frac{dC_1}{dt} = C_2 \exp(-i\omega_0 t)\mathbf{E}_0 \cdot \mu_{12} \cos(\omega t), \tag{3.15}$$

$$i\hbar \frac{dC_2}{dt} = C_1 \exp(i\omega_0 t)\mathbf{E}_0 \cdot \mu_{12} \cos(\omega t). \tag{3.16}$$

Here, μ_{12} is the dipole moment matrix element defined as

$$\mu_{ji} = \int \psi_i^* e\mathbf{r}\psi_j \, dr \ = \int \psi_i^* \mu \psi_j \, dr. \tag{3.17}$$

The matrix elements given by Equation 3.17 describe the electric dipole moments that form in the system as a result of electronic transitions between states i and j and are called *transition dipole moments* of electronic transitions. It should be noted that these are real electric dipoles that exist in the molecule only temporarily during the optical transition and that they are different from permanent dipole moments of polar molecules.

The Schrödinger equations (Equations 3.15 and 3.16) are often solved approximately using the time-dependent perturbation theory, which can be applied in cases where the Hamiltonian operator is in the form $H + H'$, and where the solutions, that is, the wave functions and energies of the unperturbed system, are known exactly. In the case of optical transitions, energies E_1 and E_2 of the ground and excited electronic states are known. At time $t = 0$, that is, before the light is turned on, there is a 100% probability that the system is in its ground state because the energy gap is too large to be overcome by thermal excitations. In Equation 3.4, this means that $C_1(0) = 1$ and $C_2(0) = 0$. After that, the light is suddenly turned on and kept on for a time interval τ. We are interested in the state of the system after this time τ. This can be obtained by integrating Equations 3.15 and 3.16 from 0 to τ and evaluating the coefficient $C_2(\tau)$. Using $C_1(0) = 1$ and $C_2(0) = 0$ gives $i\hbar(dC_1/dt) = 0$, which has a trivial solution, and

$$i\hbar \frac{dC_2}{dt} = \mathbf{E}_0 \cdot \mu_{12} \exp(i\omega_0 t) \cos(\omega t). \tag{3.18}$$

Using the identity $2\cos(\omega t) = \exp(i\omega t) + \exp(-i\omega t)$, we obtain

$$\frac{dC_2}{dt} = -i\frac{E_0 \cdot \mu_{12}}{2\hbar}(\exp(i(\omega_0 + \omega)t) + \exp(i(\omega_0 - \omega)t)). \qquad (3.19)$$

Integration of Equation 3.19 from 0 to τ gives a solution for $C_2(\tau)$ in the first order of perturbation theory as

$$C_2(\tau) = \frac{E_0 \cdot \mu_{12}}{2\hbar}\left(\frac{1 - \exp(i(\omega_0 + \omega)\tau)}{\omega_0 + \omega} + \frac{1 - \exp(i(\omega_0 - \omega)\tau)}{\omega_0 - \omega}\right). \qquad (3.20)$$

We can neglect the first term in Equation 3.20 by taking into account the magnitude of optical frequencies near resonance as $\omega_0 + \omega \gg \omega_0 - \omega$. Further, instead of the value of the coefficient $C_2(\tau)$, we will calculate the physically important quantity $C_2^*(\tau)C_2(\tau) = |C_2(\tau)|^2$, which gives the probability that a particular molecule will be in the excited state after time τ. We obtain

$$|C_2(\tau)|^2 = \frac{(E_0 \cdot \mu_{12})^2}{\hbar^2}\frac{1}{4}\left(\frac{2 - \exp(-i(\omega_0 - \omega)t) - \exp(i(\omega_0 - \omega)t)}{(\omega_0 - \omega)^2}\right), \qquad (3.21)$$

which further simplifies to

$$|C_2(\tau)|^2 = \frac{(E_0 \cdot \mu_{12})^2}{\hbar^2}\left(\frac{\sin^2(((\omega_0 - \omega)/2)\tau)}{(\omega_0 - \omega)^2}\right). \qquad (3.22)$$

The solution 3.22 is plotted in Figure 3.1 as a function of detuning, $\omega_0 - \omega$. As expected, the probability of finding a molecule in an excited

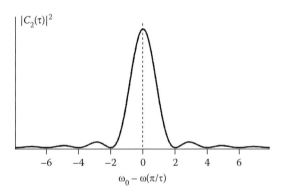

Figure 3.1 Probability of finding the system in the excited state after time τ as a function of detuning of the frequency.

state is maximum for $\omega_0 = \omega$. It can be shown that in the current approximation the probability increases as τ^2 at its maximum.

Our ultimate goal is to calculate the rate of an optical transition Γ_{12}, that is, the probability of finding the system in an excited state divided by the time interval τ. Using Equation 3.22 and the general limit

$$\lim_{\tau \to \infty} \frac{\sin^2(a\tau)}{a^2\tau} = 2\pi\delta(a), \tag{3.23}$$

we may write

$$\Gamma_{12} = \frac{\pi}{2\hbar^2}(E_0 \cdot \mu_{12})^2 \delta(\omega_0 - \omega). \tag{3.24}$$

Here, $\delta(\omega_0 - \omega)$ is Dirac's delta function. Equation 3.24 is an important result in quantum mechanics, often referred to as *Fermi's golden rule*. In its general from, Fermi's golden rule states that the probability of a quantum-mechanical transition is proportional to the square of the matrix element of the interaction Hamiltonian operator. For optical transitions, the equation shows that the probability of the transition is related to the square of the dot product of the electric field intensity and the transition dipole moment.

Equation 3.24 was derived assuming that there are two electron states in the system: the ground and excited states. In real systems, there is often a dispersion of the excited state, such as, for example, in complex organic molecules, or a continuum of states, as in the conduction bands of semi-conductors. In such cases, we replace state 2 with a number of states dn in an energy interval $d\omega_0$ in the vicinity of ω_0—that is, by DOS $\rho(\omega_0)$. To calculate the total transition rate of such excited states, we multiply the single-state rate by the DOS and integrate over the frequencies,

$$\Gamma_{1t} = \int \Gamma_{12}\rho(\omega_0)d\omega_0 = \frac{\pi}{2\hbar^2}(E_0 \cdot \mu_{12})^2 \times \int \delta(\omega_0 - \omega)\rho(\omega_0)d\omega_0$$
$$= \frac{\pi}{2\hbar^2}(E_0 \cdot \mu_{12})^2 \rho(\omega). \tag{3.25}$$

The choice of the initial conditions in Equation 3.18, that is, 100% ground-state population, means that the resulting transition rate (Equation 3.24) describes the process of *absorption* of light. Beginning from the excited state, one would be able to derive the same equation with the μ_{21} transition dipole moment, which would describe the process of *stimulated emission*. Both the processes of absorption and stimulated emission occur *only* in the presence of light. As seen from Fermi's golden rule (Equation 3.24), if $E_0 = 0$ (without light) the rate Γ_{12} is zero and transitions do not occur.

After the light is switched off, a molecule that finds itself in the excited state would stay in this state forever. This is contrary to experiments where molecules are observed to decay to their ground states spontaneously in relatively short times. This process is called *spontaneous emission*. For a full treatment of spontaneous emission, it is necessary to use the quantum theory of electromagnetic fields. In the semiclassical approach, the effect of spontaneous emission is introduced phenomenologically into the Schrödinger equation (Equation 3.16), with coefficient C_2 as an additional decay route:

$$\frac{dC_2}{dt} = -iC_1 \frac{\mathbf{E_0} \cdot \boldsymbol{\mu}_{12}}{\hbar} \exp(i\omega_0 t) \cos(\omega t) - \gamma_{SP} C_2. \tag{3.26}$$

After the light is switched off ($\mathbf{E}_0 = 0$), the equation can be solved easily to obtain

$$|C_2(t)|^2 = |C_2(0)|^2 \exp(-2\gamma_{SP} t). \tag{3.27}$$

The phenomenological constant γ_{SP} is related to the excited-state radiative lifetime τ_R as $1/\tau_R = 2\gamma_{SP}$.

Optical transitions are also described macroscopically using the Einstein coefficients of absorption B_{12}, stimulated emission B_{21}, and spontaneous emission A_{21}. For example, the absorption transition rate is given by

$$\Gamma_{12} = B_{12} W(\omega_0), \tag{3.28}$$

where $W(\omega_0)$ is the electromagnetic field energy density at a transition frequency ω_0. Comparison with Fermi's golden rule then provides the connection between the microscopic parameter of transition dipole moment and the Einstein coefficient as

$$B_{12} = |\boldsymbol{\mu}_{12}|^2 \frac{\pi}{3\varepsilon_0 \hbar^2}. \tag{3.29}$$

The relationship between the Einstein coefficients also provides a link between the transition dipole moment and the spontaneous emission rate:

$$|\boldsymbol{\mu}_{12}|^2 \frac{\pi}{3\varepsilon_0 \hbar^2} = B_{12} = \frac{\pi^2 c^3}{\hbar \omega^3} A_{12}. \tag{3.30}$$

The processes of absorption, stimulated emission, and spontaneous emission are shown symbolically in Figure 3.2.

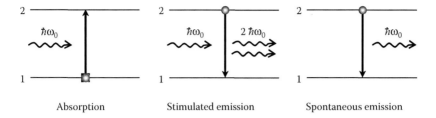

Figure 3.2 Summary of electronic optical transitions in a two-level system.

3.1.2 General concept of excitons

The term exciton has its origin in the field of solid-state physics, where it is used to describe delocalized electronic excitations in nonmetallic crystals that are characterized by band structure and a large band gap. In such systems, the exciton describes states in which an electron and a hole form a bound electron–hole pair whose energy is slightly smaller than the band-gap energy. Over the years, however, the term exciton has been used in many other areas to characterize states that do not strictly fulfill the definition as described above. Excitons are now often used to describe optical excitations in a wide range of solids, from crystals to conducting polymers and to low-dimensional organic systems. In this section, we will use the term exciton in this broader sense, even though we will also present a mathematical description of excitons based on its original meaning.

Excitons in solids are formed by two main processes: by absorption of light and by recombination of opposite charges. Other minor mechanisms include thermal release of trapped excitations or ion beam irradiation. The character and properties of excitons depend strongly on the nature of the material in which they are formed. Accordingly, excitons are classified into three main groups:

1. In materials with high dielectric constants, such as inorganic semi-conductors, the attractive Coulomb interaction between a hole and an electron are effectively screened and as a result a loosely bound exciton delocalized over many lattice sites is formed. This type of exciton is called a Wannier (or Wannier–Mott) exciton, and it is well described by the effective mass approach that results from the nearly free electron model for band structure (see Chapter 1).
2. In materials with low dielectric constants, the attractive interaction is strong and both electrons and holes are localized on the same atom or molecule in the lattice. This type of exciton is called a Frenkel exciton and is typical of organic molecular crystals. It is a direct consequence of the tight-binding model of band structure (described in Chapter 1).

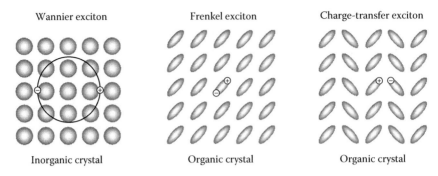

Figure 3.3 Different types of excitons classified according to the material properties.

3. A third type of exciton, which is an intermediate between Frenkel exciton and Wannier exciton, appears in ionic crystals or in molecular crystals that are formed by two kinds of molecules with different ionization potential and electron affinity. In this charge transfer (CT) exciton, electrons and holes are tightly bound and reside on different, but neighboring atoms or molecules. These three types of excitons are schematically shown in Figure 3.3.

The Wannier and Frenkel type of excitons, similar to the two models for describing solid-state band structure, can be regarded as extreme cases of a general exciton model [2]. Description of the general model is beyond the scope of this book. However, we can qualitatively illustrate the relationship between the different exciton types as well as between the exciton types and other material properties, using Figure 3.4. In the following sections, we give a detailed description of the individual exciton types and examine the effect of size—at the nanometer scale—on the exciton properties.

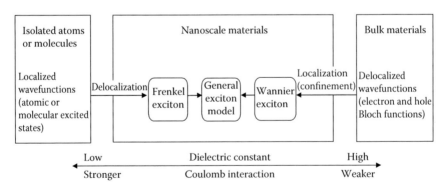

Figure 3.4 Relationship between material properties and description of excitonic states.

3.1.3 Wannier excitons

The properties of Wannier excitons in bulk material can be described fairly accurately in terms of the nearly free electron model. Electrons and holes move freely on macroscopic scales within a weak periodic potential that approximates the interaction of charge carriers with atoms in a crystal lattice. The potential energy of the charges is described by an expression resembling the energy of a free electron gas (Chapter 1), with the exception that the real masses of electrons and holes are replaced by the so-called effective masses. Thus, the change of mass alone simulates the effect of the periodic potential. The starting point in treating Wannier excitons is a Hamiltonian operator of a bound electron–hole pair, which is written in the form

$$H_{eh} = \frac{\hbar^2 k^2}{2m_e^*} + \frac{\hbar^2 k^2}{2m_h^*} - \frac{e^2}{4\pi\varepsilon \, |\mathbf{r}_e - \mathbf{r}_h|}, \tag{3.31}$$

where m_e^* and m_h^* are the electron and hole effective masses, $\mathbf{r}_e$ and $\mathbf{r}_h$ their position vectors, and ε the material permittivity. The third term on the right describes the Coulomb binding energy. Equation 3.31 is identical to the Hamiltonian operator for a hydrogen atom, and we use the known solutions of the problem introduced in Chapter 1. The energy of the excitons is described by a set of quantized energy levels $E_n(k)$:

$$E_n(k) = E_g - \frac{R_y}{n^2} + \frac{\hbar^2 k^2}{2M^*}. \tag{3.32}$$

Here, E_g is the energy of the band gap, $M^* = m_e^* + m_h^*$, and R_y is a so-called exciton Rydberg energy given by

$$R_y = \frac{\mu e^4}{2\hbar^2} \frac{1}{(4\pi\varepsilon)^2} = \frac{e^2}{2a_B} \frac{1}{4\pi\varepsilon}. \tag{3.33}$$

The term R_y/n^2 in Equation 3.32 is identical to the energy of the hydrogen atom derived in Chapter 1. In a similar manner as the quantity a_0, which described the radius of the hydrogen atom, the exciton Bohr radius a_B is given by

$$a_B = 4\pi\varepsilon \frac{\hbar^2}{\mu e^2}. \tag{3.34}$$

This quantity is very important in describing the effect of size on excitons in nanoscale systems. The μ in the above equation refers to the

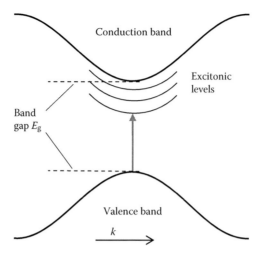

Figure 3.5 Optical transition to excitonic states in bulk semiconductors.

reduced mass defined as $\mu^{-1} = m_e^{*-1} + m_h^{*-1}$. Figure 3.5 schematically shows optical transitions to bound exciton states, as well as direct interband transitions, in a bulk semiconductor with a so-called direct band gap.

3.1.4 *Size effects in high-dielectric-constant materials*

We now turn our attention to changes of the optical excitations of materials with a high dielectric constant if their size is reduced to nanometer scales. In bulk samples, optical excitations can be either due to the formation of bound Wannier excitons with discrete energy levels, as described above, or due to direct transitions from the valence band to a continuum of states in the conduction band—that is, direct interband transitions. Decreases in size affect both types of transitions. In the case of bound exciton transitions, the size effect can be understood from Equation 3.33, which shows that the Rydberg exciton energy increases with the inverse of the Bohr radius. The size effect on excitons starts to become prominent when the physical size of a sample of material is comparable or smaller than the Bohr radius of the material. For most inorganic semiconductors, this size range is around 10 nm. Table 3.1 summarizes the Bohr radii together with band gap energies and exciton binding energies for some well-known semiconductors.

The second type of size effect is related to interband transition and is due to quantum confinement and the resulting quantization of the states at band edges, as described in detail in Chapter 2 and summarized in Figure 3.6. In a 2D confined system, that is, a thin slab of thickness l_z (nanosheet or a quantum well), electrons can move freely along the

Table 3.1 Optical Properties of Selected Semiconducting Materials

Semiconductor Material	Classification	Band Gap Transition Wavelength (nm)	Band Gap Energy (eV)	Exciton Bohr Radius (nm)
CuCl	I–VII	360	3.395	0.7
CdS	II–VI	480	2.583	2.8
CdSe	II–VI	670	1.89	4.9
GaN	III–V	360	3.42	2.8
GaP	III–V	550	2.26	10–6.5
InP	III–V	920	1.35	11.3
GaAs	III–V	870	1.42	12.5
AlAs	III–V	570	2.16	4.2
AlN	III–V	200	6.026	1.96
Si	IV	1150	1.11	4.3
Ge	IV	1880	0.66	25
PbS	IV–VI	3000	0.41	18

x- and *y*-directions and are confined along the *z*-direction. The energy of electrons is quantized and given by

$$E_{2D}(k) = \frac{\hbar^2}{2m_e^*}(k_x^2 + k_y^2) + \frac{n_z^2 h^2}{8m_e^* l_z^2}. \tag{3.35}$$

The second term on the right describes the quantization of electron energy with a quantum number n_z along the *z*-direction, where it behaves like a particle in a box (see Chapter 1). The energy of holes in such system

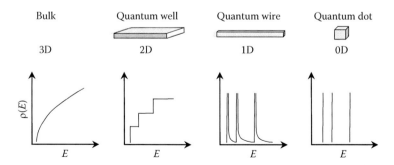

Figure 3.6 Schematic illustration of the electron DOS confined in 2D, 1D, and 0D structures.

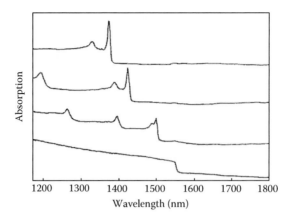

Wavelength (nm)

Figure 3.7 Absorption spectra of quantum wells of the semiconductor structure InGaAs/InAlAs with different values of the well thickness *l*. From bottom to top: 600 nm, 20 nm, 10 nm, 7.5 nm. (Reprinted from D.S. Chemla, *Phys. Today* 46, 1993, 46. With permission.)

can be described analogously. In 2D systems, the energy of the interband optical transition E_B, which is equal to the band gap energy E_g in bulk samples, changes to

$$E_B = E_g + \frac{h^2}{8\mu l_z^2}. \tag{3.36}$$

The energy of the transition increases with the inverse of the square of the dimension l_z. Figure 3.7 illustrates the size effect on both excitonic and interband transitions for quantum wells of semiconductor hetero-structure, InGaAs/InAlAs. As the size l_z decreases from 600 nm to 20, 10, and 7.5 nm, the lowest energy peak in the absorption spectrum—which corresponds to the bound exciton—shifts to higher energies. The inter-band transitions (shifted to higher energies from the exciton peak) in the "bulk" 600 nm sample show a continuous band due to the continuous DOS. This continuous band changes to a series of broad peaks in the 20 nm sample due to the quantization effect. In the 10 nm sample, these peaks further shift to higher energies. Note the different amounts of shift of the excitonic and interband transitions with decreasing l_z, due to the different scaling of excitonic and interband transitions predicted by Equations 3.33 and 3.36.

Apart from affecting the energy of optical transitions, the size effect is also apparent in changes of the strength of interband transitions. This is a consequence of Fermi's golden rule (Equation 3.25), which relates the rate of transitions to the DOS. This is demonstrated in Figure 3.7 as an increase in absorption around the peaks in the 20 and 10 nm samples due to stepwise DOS (see Chapter 2) compared with the continuous absorption band in the 600 nm sample, which is due to a continuum of states in the conduction band of the bulk sample.

In analogy with 2D systems, the size effects in 1D systems (quantum wires) can be obtained by considering the quantized energy of electrons, which move freely along the y-direction and are confined in the x- and z-directions by the dimensions l_x and l_z:

$$E_{1D}(k) = \frac{\hbar^2 k_y^2}{2m_e^*} + \frac{n_x^2 h^2}{8m_e^* l_x^2} + \frac{n_z^2 h^2}{8m_e^* l_z^2}, \tag{3.37}$$

and the energy of the interband transition is

$$E_B = E_g + \frac{h^2}{8\mu}\left(\frac{n_x^2}{l_x^2} + \frac{n_z^2}{l_z^2}\right). \tag{3.38}$$

The intensity of the transitions is modified by the DOS function of quantum wires (see Chapter 2).

Finally, in the case of a 0D system (quantum dot), electrons are confined in all three directions by the sizes l_x, l_y, and l_z. The intraband transition energy is now

$$E_B = E_g + \frac{h^2}{8\mu}\left(\frac{n_x^2}{l_x^2} + \frac{n_y^2}{l_y^2} + \frac{n_z^2}{l_z^2}\right). \tag{3.39}$$

The effect of changed DOS is most pronounced for quantum dots because (as seen in Figure 3.6) the DOS function is now a sum of sharp peaks (delta functions). All states that were continuously spread in the conduction band in the bulk state are now squeezed into a few peaks, and consequently, there is a dramatic increase in the transition rate for these peaks. In this respect, the optical properties of semiconductor quantum dots resemble those of isolated organic molecules. The large Bohr radius a_B (in the order of 10 nm) of Wannier excitons in semiconductors means that it is relatively easy to actually prepare nanoparticles (nanocrystals) with sizes $l < a_B$, both by top-down (physical) and bottom-up (chemical) approaches. By controlling the sizes of the quantum dots, it is possible to control the energies of optical transitions and consequently to tune absorption and emission wavelengths of light. Combining different materials and sizes, it is possible to cover the whole visible spectrum of light, and even go into the adjacent near-UV and infrared (IR) regions. Figure 3.8 shows some examples of absorption and emission spectra of nanocrystals of different materials and a graphical summary of the experimentally observed relationship between the crystal size and transition energy.

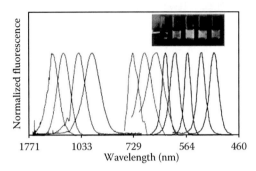
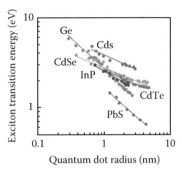

Figure 3.8 **(See color insert following page 148.)** Left: Size- and material-dependent fluorescence spectra of semiconductor nanocrystals. Blue lines: CdSe nanocrystals with diameters of 2.1, 2.4, 3.1, 3.6, and 4.6 nm (from right to left). Green lines: InP nanocrystals with diameters of 3.0, 3.5, and 4.6 nm. Red lines: InAs nanocrystals with diameters of 2.8, 3.6, 4.6, and 6.0 nm. Inset: images of a series of silica-coated core (CdSe) shell (ZnS or CdS) nanocrystals. (Reprinted from M. Bruchez et al., *Science* 281, 1998, 2013. With permission.) Right: Size-dependent exciton energies for quantum dots of various semiconducting materials. (Reprinted from G.D. Scholes and G. Rumbles, *Nat. Mater.* 5, 2006, 683. With permission.)

3.1.5   Size effects in π-conjugated systems

Conjugated organic molecules contain alternate single and double (or triple) carbon–carbon bonds. Overlap between p-type atomic orbitals causes delocalization of the ground-state electrons over the length of the conjugated system. Electrons move in a periodic potential similar to the motion of free electrons in the conduction bands of semiconductors. A rigorous theoretical treatment of conjugated organic molecules involves quantum chemical molecular orbital methods where the wave functions of the system (π-orbitals) are constructed as linear combinations of the wave functions (p-orbitals) of the atoms constituting the conjugated system. The possible linear combinations of the atomic orbitals are restricted by the requirement that the resulting molecular electronic wave functions conform to the symmetry of the molecules. To illustrate the size effect on the optical properties of conjugated systems, however, it is possible to use very simple approximate methods, such as the free electron method for linear conjugated systems or perimeter-free electron orbital (PFEO) model for aromatic systems. In both methods, it is assumed that electrons move freely along a line or a loop (in the case of PFEO), and from the quantum-mechanical point of view, the problem is identical to that of a particle in a 1D box (see Chapter 1), and the two methods differ only by the boundary conditions used. Below, we will illustrate the size effect using the solutions of the PFEO method.

We are looking for solutions of the time-independent Schrödinger equation:

$$-\frac{\hbar^2}{2m}\frac{\partial^2\psi(x)}{\partial x^2} = E\psi(x), \tag{3.40}$$

such that the eigenstates satisfy the periodic boundary condition $\psi(x) = \psi(x + l)$, where l is the length of the perimeter of the loop and x is a distance along the loop. Solutions of the Schrödinger equation give

$$\psi_0 = \sqrt{\frac{1}{l}},$$

$$\psi_{n1}(x) = \sqrt{\frac{1}{l}}\cos\left(\frac{2\pi nx}{l}\right), \tag{3.41}$$

$$\psi_{n2}(x) = \sqrt{\frac{1}{l}}\sin\left(\frac{2\pi nx}{l}\right).$$

The states with quantum number $n > 0$ are doubly degenerate because of the periodic boundary conditions. Physically, the degeneracy reflects the fact that the electron can move clockwise or anticlockwise around the perimeter. In analogy with Equation 1.39, we obtain for the energy of electron in state n:

$$E_n = \frac{n^2 h^2}{2ml^2}, \tag{3.42}$$

and for the energy of an optical transition between states n, $n + 1$:

$$E_{n,n+1} = \frac{h^2}{2ml^2}(2n+1). \tag{3.43}$$

Equation 3.43 shows the dependence of the transition energy on the perimeter length, that is, on the length of the conjugated system, as l^{-2}. With increasing conjugation length, that is, with increasing numbers of aromatic rings, the energies of the states, as well as the energy separation between adjacent states, decrease. This decrease gives rise to the well-known optical property of π-conjugated systems—the spectral red shift with increasing conjugation length. This effect is illustrated in the absorption spectra of a series of acenes in Figure 3.9.

Conjugated polymers are another important class of π-conjugated systems. Conjugated polymers typically consist of hundreds to thousands

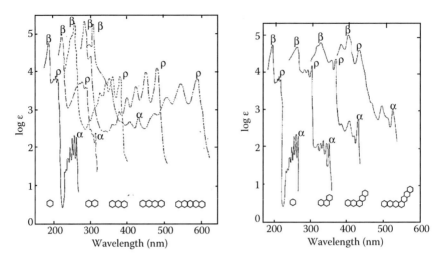

Figure 3.9 Absorption spectra of a series of acenes (left) and symmetric phenes (right) with increasing numbers of aromatic rings. (Reproduced from J.B. Birks, *Photophysics of Aromatic Molecules*, Wiley-Interscience, London, 1970. With permission.)

of monomer repeat units. A straightforward application of the free electron model would shift the transition energies out of the visible region toward the IR. However, this is not what is experimentally observed. The electrons in the real systems are delocalized over 5–20 monomer units, forming the so-called conjugated segments. The reasons for this partial localization are structural (topological) or chemical inhomogeneities or interactions of the electrons with phonons. The length of the conjugated segment can be determined by measuring the transition energies of a series of oligomers with increasing numbers of monomers. The increasing conjugation length causes a red spectral shift of the absorption or emission maxima, which saturates at oligomer sizes corresponding to the length of the conjugated segment. The effect of the oligomer length on the optical spectra of oligothiophenes and oligophenylenevinylenes is shown in Figure 3.10.

The kinds of inhomogeneities that interrupt the electron delocalization in conjugated polymers are due to the amorphous nature of the solid state of the flexible or semirigid polymers. These inhomogeneities are almost absent in another important class of carbon-based material—the carbon nanotubes. The electrons in these systems exhibit much larger delocalization, and the corresponding optical transition energies lie in the near-IR spectral region. Figure 3.11 summarizes the experimentally observed relationship between the size and transition energy of organic systems.

It is interesting to note that the size dependence of optical transitions of both semiconducting nanostructures (quantum dots) and π-conjugated

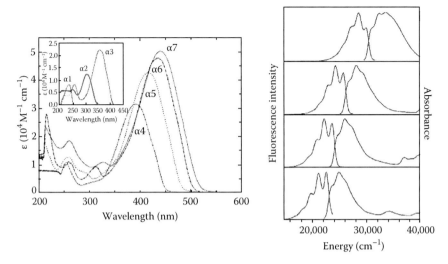

Figure 3.10 Size effects in the optical properties of conjugated oligomers. Left: Absorption spectra of a series of oligothiophenes for the number of monomers α from 1 to 7. (Reprinted from R.S. Becker et al., *J. Phys. Chem.* 100, 1996, 18683. With permission.) Right: Absorption and fluorescence spectra of a series of oligophenylenevinylenes for the number of monomers from 1 to 4. (Reprinted from J. Gierschner et al., *J. Chem. Phys.* 116, 2002, 8596. With permission.)

organic systems is well described by the simple quantum particle-in-a-box model, cf Equations 3.39 and 3.43. However, the solutions were reached from opposite directions: downsizing a bulk material to below the exciton Bohr radius in semiconductors and building up a delocalized π-conjugated system in organic molecules.

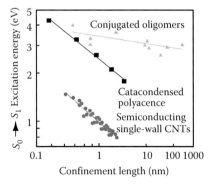

Figure 3.11 Size-dependent transition energies of a selection of π-conjugated organic systems obtained from absorption spectra. (Reprinted from G.D. Scholes and G. Rumbles *Nat. Mater.* 5, 2006, 683. With permission.)

3.1.6 Strongly interacting π-conjugated systems: A molecular dimer

A molecular dimer is a pair of identical molecules that are bound in their ground electronic states by weak intermolecular forces (van der Waals forces, hydrogen bonds). The resulting close proximity of the molecules then leads to a strong interaction on optical transitions. Molecular dimers are the basic building block for larger-scale structures, from linear molecular aggregates to molecular crystals, and the optical properties of these systems are a direct extension of those of the dimers. Optical transitions also show strong distance dependence. As such, we describe dimers in considerable detail and use the results in later sections for larger-scale systems.

The theoretical treatment [10] is based on an assumption that electron orbital overlap and electron exchange between two molecules of a dimer are negligible. Electrons responsible for optical transitions are localized on the constituent molecules, and the molecular units of the system retain their individual characteristics. Analysis of the electron states of complexes can thus proceed in terms of the electron states of individual isolated molecules. Note that this treatment follows the same direction as the tight-binding model for band structure introduced in Chapter 1. We will consider only two electron levels of the constituent monomer molecules. Their ground-state wave functions are ψ_1 and ψ_2, where the numerical subscripts refer to molecules 1 and 2. The wave functions of excited states are written with the superscript u (upper) as ψ_1^u and ψ_2^u. The ground-state wave function of a dimer is a product of the monomer wave functions:

$$\Psi_G = \psi_1 \psi_2, \tag{3.44}$$

and the excited-state wave function is written as a linear combination of two possible nonstationary excited states either with molecule 1 or with molecule 2 excited:

$$\Psi_E = a\psi_1^u \psi_2 + b\psi_1 \psi_2^u. \tag{3.45}$$

The coefficients a and b are determined later. The Hamiltonian operator of the dimer is

$$H = H_1 + H_2 + V_{12}, \tag{3.46}$$

where H_1 and H_2 are Hamiltonian operators for the isolated monomers and V_{12} is an operator for the intermolecular interaction. Using the Hamiltonian operator 3.46 in the time-independent Schrödinger equation for the dimer ground state:

$$H\Psi_G = E_G \Psi_G, \tag{3.47}$$

we obtain for the dimer ground-state energy:

$$E_G = \iint \psi_1^* \psi_2^* H \psi_1 \psi_2 \, dr_1 \, dr_2 = E_1 + E_2 + \iint \psi_1^* \psi_2^* V_{12} \psi_1 \psi_2 \, dr_1 \, dr_2$$
$$= E_1 + E_2 + D_G. \tag{3.48}$$

The terms E_1 and E_2 are the ground-state energies of the isolated monomers and D_G is the energy correction due to van der Waals interaction in the ground state, lowering the E_G. Writing the Schrödinger equation for the excited state, we obtain

$$H(a\psi_1^u \psi_2 + b\psi_1 \psi_2^u) = E_E(a\psi_1^u \psi_2 + b\psi_1 \psi_2^u). \tag{3.49}$$

Multiplying Equation 3.49 from the left alternately by $\psi_1^{u*} \psi_2^*$ and $\psi_1^* \psi_2^{u*}$ and integrating each time over coordinates of both molecules lead to a set of equations:

$$aH_{11} + bH_{12} = aE_E,$$
$$aH_{21} + bH_{22} = bE_E, \tag{3.50}$$

where

$$H_{11} = \iint \psi_1^{u*} \psi_2^* \psi_1^u \psi_2 \, dr_1 \, dr_2 \quad \text{and} \quad H_{12} = \iint \psi_1^{u*} \psi_2^* H \psi_1 \psi_2^u \, dr_1 \, dr_2. \tag{3.51}$$

It follows from the symmetry of the problem that $H_{11} = H_{22}$ and $H_{12} = H_{21}$. Nontrivial solutions of the set of Equation 3.50 are found by the requirement that the determinant of the set must be zero:

$$\begin{vmatrix} H_{11} - E_E & H_{12} \\ H_{12} & H_{11} - E_E \end{vmatrix} = 0. \tag{3.52}$$

This gives two solutions for the excited energy states in the form of

$$E_E' = H_{11} + H_{12} \quad \text{and} \quad E_E'' = H_{11} - H_{12}, \tag{3.53}$$

and two corresponding wave functions:

$$\Psi_E' = \frac{1}{\sqrt{2}(\psi_1^u \psi_2 + \psi_1 \psi_2^u)} \quad \text{and} \quad \Psi_E'' = \frac{1}{\sqrt{2}(\psi_1^u \psi_2 - \psi_1 \psi_2^u)}. \tag{3.54}$$

Using the Hamiltonian operator 3.46 in Equation 3.53, we obtain

$$E'_E = E_1^u + E_2 + \iint \psi_1^{u*}\psi_2^{*}V_{12}\psi_1^u\psi_2 \, d r_1 \, d r_2 + \iint \psi_1^{u*}\psi_2^{*}V_{12}\psi_1\psi_2^u \, d r_1 \, d r_2,$$

$$E''_E = E_1^u + E_2 + \iint \psi_1^{u*}\psi_2^{*}V_{12}\psi_1^u\psi_2 \, d r_1 \, d r_2 - \iint \psi_1^{u*}\psi_2^{*}V_{12}\psi_1\psi_2^u \, d r_1 \, d r_2.$$

$$(3.55)$$

The first integral on the right in the equations represents the van der Waals interaction between an excited state of molecule 1 and the ground state of molecule 2. The last term in both equations is called the exciton displacement energy, the exciton splitting energy, or the resonance interaction energy. It describes the exchange of excitation between molecules 1 and 2. Equation 3.55 can be rewritten as

$$E'_E = E_1^u + E_2 + D_E + \varepsilon \quad \text{and} \quad E''_E = E_1^u + E_2 + D_E - \varepsilon. \qquad (3.56)$$

The interaction between the two molecules in the dimer leads to a lowering of their combined ground-state energy due to van der Waals interaction and lowering and splitting of their excited-state energies due to interaction with light. These effects are schematically shown in Figure 3.12. In terms of the optical transition energies between the excited and ground states, we can write

$$E_E - E_G = \Delta E = \Delta E_{\text{monomer}} + \Delta D \pm \varepsilon. \qquad (3.57)$$

To proceed further, we must specify the form of the interaction Hamiltonian operator V_{12}. Generally, this operator includes Coulomb interaction between all electrons and nuclei of the interacting molecules. The relevant terms cannot be easily evaluated by such an operator. Instead, the operator is usually expanded into a series of multipoles, containing

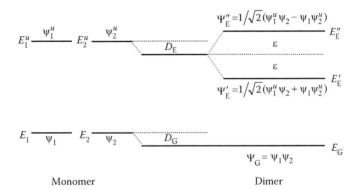

Figure 3.12 Change of energy levels of a monomer molecule upon interaction in a dimer.

monopole–monopole, monopole–dipole, dipole–dipole, quadrupole–quadrupole, and higher interactions. For neutral molecules interactions involving monopoles are zero, and for allowed optical transitions, the dipole–dipole interaction is the dominating term. In the dipole–dipole approximation, the Hamiltonian operator V_{12} is determined by the potential energy of two dipoles $\mathbf{p}_1$ and $\mathbf{p}_2$ of the two molecules. In classical electromagnetic theory, the energy of two dipoles randomly oriented in space and separated by a distance r_{12} is expressed as

$$V_{12} = \frac{1}{4\pi\varepsilon_0}\left(\frac{\mathbf{p}_1 \cdot \mathbf{p}_2}{r_{12}^3} - \frac{3(\mathbf{p}_1 \cdot \mathbf{r}_{12})(\mathbf{p}_2 \cdot \mathbf{r}_{12})}{r_{12}^5} \right), \tag{3.58}$$

where $\mathbf{r}_{12}$ is a vector connecting the centers of the molecules. To simplify the analysis, we assume that the molecules are oriented parallel to each other (face-to-face) as shown in Figure 3.13. The vectors of the dipole moments $\mathbf{p}_1$ and $\mathbf{p}_2$ lie in the plane of the molecules and we write $\mathbf{p}_1 = e\mathbf{r}_1$ and $\mathbf{p}_2 = e\mathbf{r}_2$, where the vectors $\mathbf{r}_1$ and $\mathbf{r}_2$ are in the x–y plane. The interaction term then simplifies to

$$V_{12} = \frac{e^2}{4\pi\varepsilon_0}\left(\frac{\mathbf{r}_1 \cdot \mathbf{r}_2}{r_{12}^3} \right). \tag{3.59}$$

Using this expression for the interaction Hamiltonian operator, we may now evaluate the exciton displacement term ε in Equation 3.55:

$$\varepsilon = \iint \psi_1^{u*}\psi_2^*V_{12}\psi_1\psi_2^u \, dr_1 \, dr_2$$

$$= \frac{1}{4\pi\varepsilon_0 r_{12}^3}\left(\int \psi_1^{u*}e\mathbf{r}_1\psi_1 \, dr_1 \right) \cdot \left(\int \psi_2^{u*}e\mathbf{r}_2\psi_2 \, dr_2 \right). \tag{3.60}$$

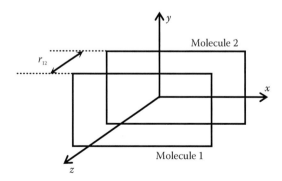

Figure 3.13 Molecular arrangement in a parallel dimer.

In the integrals on the right of Equation 3.56, we recognize the transition dipole moments of individual monomer molecules as defined in Equation 3.17. Using μ_1 and μ_2 for the transition dipole moments of molecules 1 and 2, we write

$$\varepsilon = \frac{\mu_1 \cdot \mu_2}{4\pi\varepsilon_0 r_{12}^3}. \tag{3.61}$$

Here, we arrive at an important result, namely, the resonant interaction is determined by the transition dipole moments of the two molecules and that it scales as the inverse of the third power of the intermolecular distance.

Although the monomer excited state is split into two by the exciton splitting, both of the dimer excited states may not necessarily take part in an optical transition. The probability of the transition into the Ψ'_E and Ψ''_E states is given by the magnitude of the total transition dipole moments $\mathbf{M'}$ and $\mathbf{M''}$, which are expressed as

$$\mathbf{M'} = \iint \Psi_G(\hat{\mu}_1 + \hat{\mu}_2)\Psi'_E \, dr_1 \, dr_2 = \frac{1}{\sqrt{2}}(\mu_1 + \mu_2),$$

$$\mathbf{M''} = \iint \Psi_G(\hat{\mu}_1 + \hat{\mu}_2)\Psi''_E \, dr_1 \, dr_2 = \frac{1}{\sqrt{2}}(\mu_1 - \mu_2). \tag{3.62}$$

Here, $\hat{\mu}_1$ and $\hat{\mu}_2$ are transition dipole moment operators for molecules 1 and 2. The actual values of $\mathbf{M'}$ and $\mathbf{M''}$ depend on the concrete geometry of the dimer. For the parallel dimer in Figure 3.13, a simple consideration of the total energy of a pair of dipoles implies that the lower energy state is the one with opposite orientation of the transition dipole moments. Since the molecules are identical, $|\mu_1| = |\mu_2| = |\mu|$ and we obtain from Equation 3.62 $\mathbf{M'} = 0$ and $\mathbf{M''} = \sqrt{2}\mu$. For a parallel dimer, the optical transition occurs only to the upper excited state Ψ''_E. This simple vector consideration can be generalized to predict the relative energies, phases of excited-state wave functions, and optical transition strengths of different dimer geometries [10]. The situation for dimers in parallel, head-to-tail, and oblique geometries is depicted in Figure 3.14. After absorption, the energy relaxes nonradiatively to the lowest excited state. In head-to-tail and oblique geometries, the transitions between the lowest excited and the ground states are allowed and such dimers emit light—fluoresce. However, this is not the case for parallel dimers where the transition to the lowest excited state is forbidden. Consequently, parallel dimers of even strongly fluorescent molecules are nonemissive.

The distance dependence of the splitting of the excited-state energy predicted in Equation 3.61 has been experimentally observed for a series of naphthyl dimers covalently linked via a stiff spacer of increasing length

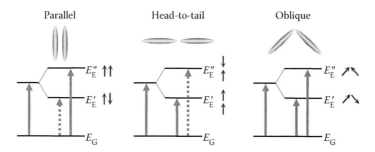

Figure 3.14 Absorption transition probabilities in dimers of different geometrical arrangements.

(Figure 3.15). In dimers, the transition dipoles of the naphthyls correspond to the oblique orientation, and as a result, the monomer absorption band splits into two in dimers because of the nonzero transition probability of both excited states. The splitting of the absorption peaks in oblique dimers is given by 2ε and is called the Davidov splitting. With decreasing spacer length, the ε increases and the separation of the two absorption peaks increases.

3.1.7 Molecular Frenkel exciton

The theoretical approach developed for molecular dimers can be easily extended to the concept of 1D linear chain of molecules [10]. Such chains are a good model for linear molecular aggregates or for linear 1D molecular crystals with only one molecule in each unit cell. Because the chains are assumed to be of infinite lengths, the model does not require boundary conditions. This fact makes analysis easy, and the model best describes

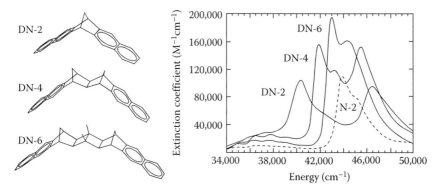

Figure 3.15 Molecular structure (left) and absorption spectra (right) of a naphthyl dimer covalently linked via a spacer with increasing length. (Reprinted from G.D. Scholes et al., *J. Am. Chem. Soc.* 115, 1993, 4345. With permission.)

the case for very long chains. However, this approach is not accurate for systems of intermediate length (3–10 molecules). We will assume a linear chain composed of N identical molecules (where N is very large). In analogy with the dimer analysis, the ground-state wave function is a product of the ground-state wave functions of the individual molecules:

$$\Psi_G = \psi_1 \psi_2 \psi_3 \cdots \psi_N = \prod_{n=1}^{N} \psi_n. \tag{3.63}$$

The excited state can be written as

$$\Phi_a = \psi_1 \psi_2 \psi_3 \cdots \psi_a^u \cdots \psi_N = \psi_a^u \prod_{\substack{n=1 \\ n\neq a}}^{N} \psi_n. \tag{3.64}$$

This is an excited state of the aggregate, where the monomer molecule a is in the excited state and other molecules are in their ground states. There are N such products and they represent nonstationary excited states that are not eigenfunctions of the system Hamiltonian operator. The excited state of the system is written as a linear combination of the states expressed by Equation 3.64:

$$\Psi_E(k) = \sum_{a=1}^{N} C_a(k)\Phi_a. \tag{3.65}$$

The variable k describes the kth excited state. Using the concept of Bloch functions (Chapter 1), Equation 3.65 can be expressed as

$$\Psi_E(k) = \frac{1}{\sqrt{N}} \sum_{a=1}^{N} \Phi_a \exp\left(\frac{i2\pi ka}{N}\right), \tag{3.66}$$

or in terms of the wavevector $\mathbf{k}$ and position vector $\mathbf{r}_a$ of the molecule a as

$$\Psi_E(\mathbf{k}) = \frac{1}{\sqrt{N}} \sum_{a=1}^{N} \Phi_a \exp(i\mathbf{k}\cdot\mathbf{r}_a). \tag{3.67}$$

Equations 3.66 and 3.67 describe the excited state of the system as a linear combination of excited states of individual monomer molecules. The effect of the linear combination is that the wave function is one for a collective excitation of N molecules in the chain. The excitation is delocalized over N monomer molecules and possesses well-defined phases for each monomer. The oscillating term in Equations 3.66 and

3.67 implies that the excitation can be characterized by a wavelength λ defined by the wavevector as $|\mathbf{k}| = 2\pi/\lambda$ and by a momentum $\mathbf{p} = \hbar\mathbf{k}$. This type of excitation is called a molecular exciton. To distinguish it from the concept of Wannier excitons used in semiconductor crystals, this type of exciton is called the Frenkel exciton. The main difference from Wannier excitons is that in Frenkel excitons the excited-state electrons (or electrons and holes) are localized on individual monomer molecules. The delocalization refers to the excitation that is spread over N molecules. The delocalization length (or equivalently the number N) is called the *coherent length* of the exciton.

By adopting the procedure used to determine the energies of the excited states of dimers, it is possible to use the wave functions 3.63 and 3.66 to calculate the energy levels of the system. The interaction potential is again approximated by dipole–dipole interactions, and as a further simplification, the interaction is considered to take place only between nearest neighbors. The exciton displacement term of the interaction potential is then expressed as

$$\varepsilon_{a,a+1} = \int \Phi_a^* V_{a,a+1} \Phi_{a+1}\, dr,\tag{3.68}$$

which describes the transfer (displacement) of the excitation from a state with molecule a excited to a state with molecule $a + 1$ excited, that is, transfer of excited energy between neighboring molecules. For large N, the exciton state energies are given by

$$E_E(k) = E_{E,a} + 2\varepsilon \cos\left(\frac{2\pi k}{N}\right),\tag{3.69}$$

where $E_{E,a}$ is the energy of the excited state of molecule a and $k = 0, \pm1, \pm2, \ldots, N/2$. The subscripts in ε are omitted because all paired molecular interactions are equivalent and the term D_E in Equation 3.56 is neglected. The energy of the excited state of the monomer molecule is now split into k levels ranging from $E_{E,a} - 2\varepsilon$ to $E_{E,a} + 2\varepsilon$. These levels form a quasicontinuous energy band of width 4ε.

The actual form of the dipole–dipole interaction term depends on the geometry of the problem. For a linear chain of molecules inclined at an angle α from the chain axis and with a center-to-center distance d (see Figure 3.16), the excited-state energy levels can be written as a function of the angle α and, using the magnitude of the monomer transition dipole moment μ, in the form

$$E_E(k) = E_{E,a} - 2\left(\frac{\mu^2}{4\pi\varepsilon_0 d^3}\right)(1 - 3\cos^2\alpha)\cos\left(\frac{2\pi k}{N}\right).\tag{3.70}$$

Figure 3.16 Linear chain of molecules forming an aggregate.

Using the terminology for molecular aggregates, the case for $\alpha = 90°$ corresponds to the face-to-face or card-pack arrangement of molecules, often called the H-type aggregate. The case of $\alpha = 0°$ is the head-to-tail configuration, also called the J-type aggregate. The strength of the optical transitions can be found using the transition moment vector approach that was used for molecular dimers. In analogy to molecular dimers, the aggregate transition dipole moment of allowed transitions is $\mathbf{M} = \sqrt{N}\mu$. For H-aggregates, the transition to the lowest exciton level is forbidden and for J-aggregates this transition is allowed. All intermediate transitions for these two types are forbidden. The energy-level scheme and the optical transitions for linear molecular aggregates are summarized in Figure 3.17.

Next, let us consider a 3D molecular crystal with a simple cubic lattice, with one molecule in each unit cell and with a lattice constant d.

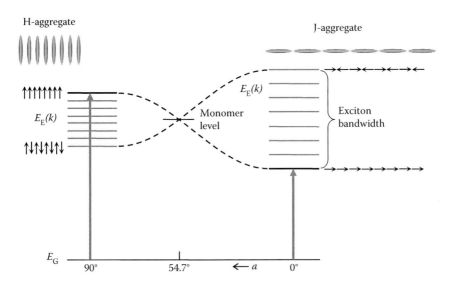

Figure 3.17 Energy levels and transition probabilities in a linear chain aggregate.

Neglecting again the energy D_E, the exciton energy can be written as (see also Chapter 1)

$$E(\mathbf{k}) = E_{E,a} - 2\varepsilon(\cos k_x d + \cos k_y d + \cos k_z d). \tag{3.71}$$

When compared with the 1D molecular aggregate, the bandwidth in a 3D molecular crystal increases to 12ε. For the general case of a crystal with two molecules in a unit cell, the situation is more complicated because we have to distinguish between two types of dipole–dipole interactions. Let us assume that molecule *a* will be residing in unit cell α and molecule *b* in a unit cell β. Taking into account all interactions, the energy of the exciton is written as (see Chapter 1)

$$E_E(\mathbf{k}) = E_{E,a} + \sum_a \varepsilon_{\alpha a, \beta a} \exp(i\mathbf{k} \cdot (\mathbf{R}_{\alpha a} - \mathbf{R}_{\beta,a})) \\ + \sum_a \varepsilon_{\alpha a, \beta b} \exp(i\mathbf{k} \cdot (\mathbf{R}_{\alpha a} - \mathbf{R}_{\beta,b})). \tag{3.72}$$

Here, the second term corresponds to translationally equivalent interactions and the third term to translationally inequivalent interactions.

3.1.8 Size effects in molecular excitons: Coherence length and cooperative phenomena

In molecular excitons, both electrons and holes are localized on the same molecule. The exciton radii are very small compared with Wannier excitons, and the effects of quantum confinement that are prominent in semiconductor crystals are absent in molecular crystals or in materials of low dielectric constant. However, Frenkel excitons in many systems are delocalized over large distances, that is, they posses large coherence length N. In such systems, the absorption and emission processes occur in phase for large numbers of molecules and can be considered as being cooperative phenomena. At the same time, the strength of allowed optical transitions, that is, transition rates, is given by the square of the transition dipole moment $\mathbf{M}^2$ (consequence of Fermi's golden rule, Equation 3.24), which is determined by the monomer dipole moments μ as $\mathbf{M}^2 = N\mu^2$. The resulting giant transition strength gives rise to new photophysical phenomena. As a result, considerable size effects can be observed as a result of changes of the coherence length.

One consequence of large coherent length is the effect of motional narrowing of the absorption and fluorescence spectra. An example of narrowing of the absorption spectrum for the J-aggregate of cyanine dyes is shown in Figure 3.18. The large width of the monomer absorption band

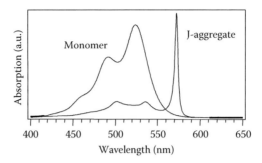

Figure 3.18 Absorption spectra of pseudoisocyanine chloride dye in monomer and J-aggregate form.

is caused partly by static disorder of the molecules, that is, by dispersion in their optical transition energies caused by the interaction with the surrounding environment. In the J-aggregate spectrum, the dipole–dipole interaction causes a red shift of the spectrum while motional narrowing causes a decrease in the bandwidth. The narrowing is due to averaging of the static disorder in transition energies of individual molecules in the aggregate by the delocalized exciton. It can be shown theoretically that for a system where there is no correlation in disorder among individual monomers, the absorption band is narrowed by a factor of $1/\sqrt{N}$, where N is the exciton coherence length. The width of the band can be followed for small aggregates of different sizes when they are prepared by adsorption on AgBr microscrystals. Figure 3.19 shows fluorescence spectra of cyanine dye aggregates formed from different mole fractions of the dye. The increasing aggregate size corresponds to the narrowing and shift of the spectrum.

Another effect observed for delocalized excitons is the shortening of the excited-state lifetime due to cooperative emission of the N molecules of the exciton. This phenomenon is called exciton superradiance. It is caused by the fact that only one (the lowest) of the manifold of the states constituting the exciton band is participating in the emission. The transition strength of all monomer molecules is thus concentrated into this one level and the strength is given by the number of molecules N. The optical transition strength (or rate) is related via Einstein's coefficients to the radiative lifetime τ_R. Shortening of the exciton radiative lifetime τ_R is thus proportional to $1/N$ in an ideal aggregate of the coherent length N. An example of the shortening of the exciton lifetime with increasing exciton length is shown in Figure 3.20 for a model compound of monomer, dimer, and trimer of a perylene dye derivative. The lifetime change is smaller than predicted for an ideal aggregate due to effects related to disorder. For large aggregates, the coherence length does not have to correspond to the physical size of the aggregate. The loss of coherence, or dephasing, is

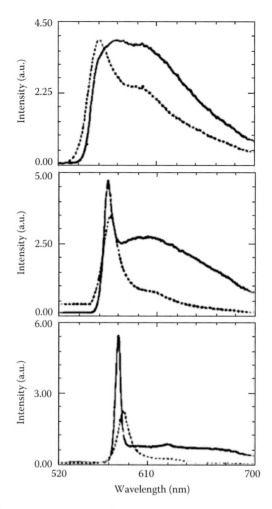

Figure 3.19 Fluorescence spectra of pseudoisocyanine (PIC) toluenesulfonate dye adsorbed on AgBr nanocrystals at different concentrations. From top to bottom: 0.125, 0.5, and 1.0 PIC mol fraction. (Reprinted from A.A. Muenter et al., *J. Phys. Chem.* 96, 1992, 2783. With permission.)

caused by collisions of excitons with phonons (vibrations of molecules or of the surrounding matrix) upon which the exciton changes its momentum (wavevector **k**) and looses coherence. The exciton–phonon interaction is strongly temperature dependent. With lowering temperature, the number of phonons decreases and exciton coherent lengths increase. This is well demonstrated in Figure 3.21 in the example of the temperature dependence of superradiant lifetime of a cyanine dye aggregate.

The $\mathbf{M} = \sqrt{N}\mu$ scaling of the aggregate transition dipole moment also leads to a strong size dependence of nonlinear optical properties.

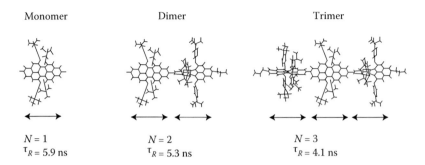

Monomer   Dimer   Trimer

$N = 1$ $N = 2$ $N = 3$
$\tau_R = 5.9$ ns $\tau_R = 5.3$ ns $\tau_R = 4.1$ ns

Figure 3.20 Monomer, dimer, and trimer of a perylene derivative and corresponding radiative lifetimes τ_R showing a decrease due to superradiance. (Data from J. Hernando et al., *Phys. Rev. Lett.* 93, 2004, 236404.)

Specifically, the third-order nonlinear susceptibility $\chi^{(3)}$, which is responsible for the phenomena of third harmonic generation or two-photon absorption, is generally proportional to $\mathbf{M}^4$. For an ideal aggregate near resonance (i.e., for the frequency of light near the frequency of the optical transition) with coherence length N, we obtain $\chi^{(3)} \propto N^2$ [15]. Note that for an ensemble of N noninteracting molecules the scaling would be linear ($\chi^{(3)} \propto N$). However, the square dependence is a theoretical prediction; in real systems, the effects of static disorder and dephasing tend to reduce the

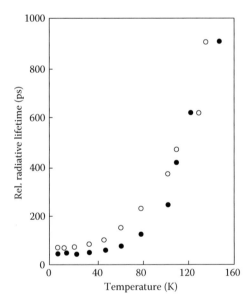

Figure 3.21 Radiative lifetime as a function of temperature for pseudoisocyanine bromide J-aggregates. (Reproduced from H. Fidder and D.A. Wiersma, *Phys. Stat. Sol. B* 188, 1995, 285. With permission.)

predicted $\chi^{(3)}$. For light frequencies far from resonance, the theory predicts no enhancement of $\chi^{(3)}$ compared with the case of N monomers [15]. Near resonance, enhanced $\chi^{(3)}$ values in the order of 10^{-6} e.s.u. for J-aggregates have been reported [16].

3.1.9 *Effects of a finite number of optical electrons*

An interesting consequence of the reduction in the size of semiconductor quantum dots below the exciton Bohr radius (i.e., below a few nanometers) is that under normal conditions only one exciton is created in the dot by absorption of light, that is, only one electron is excited from the quantized states in the valence band to the higher energy states in the conduction band. When the processes of absorption and emission are observed for individual isolated quantum dots, this fact gives rise to new photophysical phenomena that are otherwise known to occur in single atoms or molecules. One such phenomenon is related to the quantum nature of the emission of light, which can be described by Poisson statistics. After a photon has been emitted from the dot, it takes a finite time before another photon can be emitted. This is because after emitting the initial photon, the dot finds itself in the ground state, and a finite, minimum duration of time is necessary before the next process of absorption and emission. After the absorption, the dot stays in the excited state for a time interval given by the radiative (or generally fluorescence) lifetime. This whole process can be characterized by plotting the probability that a second photon is emitted at time τ after a photon has been emitted at time $\tau = 0$. This probability is given by the second-order intensity correlation function $g^{(2)}(\tau)$ in the form

$$g^{(2)}(\tau) = \frac{\langle I(t)I(t+\tau)\rangle}{\langle I(t)\rangle^2}. \qquad (3.73)$$

Here, $I(t)$ is the light intensity and $\langle\rangle$ denotes the average time over the period of the experiment. Figure 3.22 shows the probability given by $g^{(2)}(\tau)$ measured for single quantum dots of the semiconductors CdSe/ZnS. The time necessary for the emission of the next photon is demonstrated as a dip in the function at 0 times. In the field of quantum theory of light, this phenomenon is called photon antibunching. Fitting the rise in the correlation function gives the value of the dominant relaxation process, which in this case is fluorescence.

Apart from the two-electron levels considered so far (ground and excited state), there can be a third, nonemitting "trap" level present in the system of a single quantum emitter. Population of this trap level from the excited state leads to an intermission in the system fluorescence while detrapping causes fluorescence recovery. When recording fluorescence

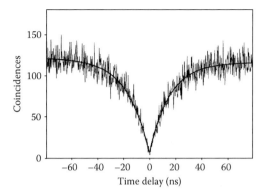

Figure 3.22 Histogram of time delays between consecutive photons detected from the fluorescence of a single CdSe/ZnS quantum dot. (Reprinted from B. Lounis et al., *Chem. Phys. Lett.* 329, 2000, 399. With permission.)

intensity as a function of time (fluorescence time trace), the steady-state intensity level is interrupted by dark periods. This phenomenon has been known as fluorescence intermittency or fluorescence blinking and is observed for virtually every single quantum emitter, from organic molecules to semiconductor quantum dots. In the case of CdSe quantum dots, the dark states are due to ionization of the dots via Auger processes, and trap states are located either on the nanocrystal surface or in the surrounding matrix. Figure 3.23 shows a time trace at different scales for a single CdSe nanocrystal. The time traces can be analyzed in terms of distributions of the light τ_{on} or dark τ_{off} time intervals. The power-law distribution of the off times, as seen in Figure 3.23, is indicative of dispersive kinetics of the detrapping process, due to the distribution of either trap depths or tunneling barriers.

Analysis of the antibunching phenomenon and fluorescence time traces is useful also in systems that consist of a small number of multiple emitters, such as collapsed chains of conjugated polymers. The different intensity levels in the intermittency in the time trace of the polymer MEH-PPV (poly2-methoxy-5-[2′-ethylhexyloxy]-1,4-phenylene vinylene) (Figure 3.24) point to simultaneous emission of a few conjugated segments. The antibunching in the correlation function is a small dip that does not reach zero, which means that the emission of individual segments in the polymer chain is uncorrelated.

3.2   *Size-dependent optical properties: Absorption and scattering*

In contrast to the processes of absorption and emission, light scattering is accurately described by analysis based on classical physics. This section

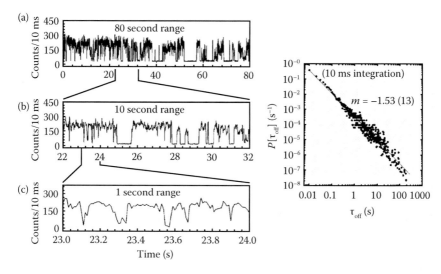

Figure 3.23 Left: Fluorescence time trace of a single ~2.9 nm radius CdSe/ZnS quantum dot showing blinking over a sequence of three expanded time scales. Right: Normalized probability densities of off times for a single ~2.9 nm radius CdSe/ZnS quantum dot. (Reprinted from M. Kuno et al., *J. Chem. Phys.* 112, 2000, 3117. With permission.)

starts with a summary of the theoretical description of size effects in light scattering. Basic scattering phenomena can be understood by considering radiation of an oscillating electric dipole represented by the Lorentz model of an atom. Size dependence of scattering is a result of rigorous solutions of Maxwell's equations for the interaction of light with dielectric spheres of arbitrary size. The solutions are known as Mie's scattering

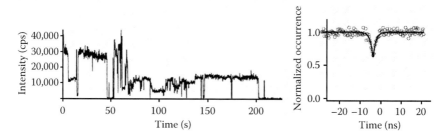

Figure 3.24 Left: Fluorescence intensity time trace of a single chain of the conjugated polymer MEH–PPV cast from toluene solution. (Reprinted from Huser, T., Yan, M., and Rothberg, L.J., *PNAS* 97, 2000, 11187. With permission.) Right: Histogram of time delays between consecutive photons in pairs detected from the fluorescence of a single MEH–PPV chain cast from toluene solution. (Reprinted from C.W. Hollars, S.M. Lane, and T. Huser, *Chem. Phys. Lett.* 370, 2003, 393. With permission.)

theory, even though in a strict sense it is not a new theory but rather a universal description of the scattering phenomenon based on Maxwell's equations. The theoretical introduction is followed by an analysis of absorption and scattering effects in small metallic nanoparticles where strong size-dependent phenomena occur on scales ranging from a few nanometer to several hundreds of nanometers. The microscopic nature of metals brings about a new phenomenon called localized plasmon resonance. The strong size dependence of surface plasmons has many potentially interesting applications, including surface-enhanced Raman scattering.

3.2.1 Basic theory of light scattering

The classical treatment of the interaction of light with matter is based on the microscopic model by H.A. Lorentz, called the Lorentz oscillator model. In the model, it is assumed that in a material medium, electrons in atoms are attached to the atomic nuclei via a classical spring, as shown in Figure 3.25, and that interaction with electromagnetic waves causes an oscillating motion of the electrons. The equation of motion of electrons based on Newton's second law of motion is

$$m\frac{d^2x}{dt^2} = F_C - F_R = eE - k_S x, \qquad (3.74)$$

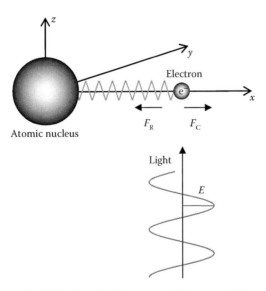

Figure 3.25 Principle of the classical Lorentz oscillator model.

where the driving force consists of a Coulomb force F_C due to the electric field and a restoring force F_R due to the spring. Here, m is the electron mass and k_S the spring constant related to the natural frequency ω_0 as $\omega_0 = \sqrt{k_s/m}$. The equation of motion in a periodic electric field of light (Equation 3.13) is a differential equation:

$$m\frac{d^2x}{dt^2} + m\omega_0^2 x = eE_0 \cos(kz - \omega t). \tag{3.75}$$

This can easily be solved by using an oscillating trial solution, which gives

$$x = \frac{e/m}{\omega_0^2 - \omega^2} E_0 \cos(kz - \omega t) = \frac{e/m}{\omega_0^2 - \omega^2} E. \tag{3.76}$$

Changes in the equilibrium positions of electrons result in an electric dipole moment p:

$$p = ex = \frac{e^2/m}{\omega_0^2 - \omega^2} E = \alpha(\omega)E. \tag{3.77}$$

The term relating the electric field and resulting dipole moment is the frequency-dependent atomic electron polarizability $\alpha(\omega)$. For an ensemble of N atoms, the individual atomic dipoles add up to create a macroscopic polarization $P = Np$. On the other hand, classical electromagnetic theory gives the macroscopic polarization in the form $\varepsilon_r = 1 + P/\varepsilon_0 E$. Combining with the Maxwell's relation $\varepsilon_r = n^2$, we can write an expression for the refractive index:

$$n^2 = 1 + \frac{N\alpha}{\varepsilon_0} = 1 + \frac{Ne^2}{m\varepsilon_0} \frac{1}{\omega_0^2 - \omega^2}. \tag{3.78}$$

This equation is valid in spectral regions far from the resonance frequency ω_0. In the vicinity of ω_0, Equation 3.78 predicts a singularity that actually does not occur. The singularity is avoided by introducing damping into the oscillator motion by adding a frictional force on the right-hand side of Equation 3.74. The frictional force is proportional to the velocity of electrons and acts along the x-axis. The equation of motion is now modified to

$$\frac{d^2x}{dt^2} + \gamma\frac{dx}{dt} + \omega_0^2 x = \frac{e}{m} E_0 e^{-i(\omega t - kz)}, \tag{3.79}$$

where we have used the complex notation for the electric field. As a result, the polarizability is now a complex quantity:

$$\alpha(\omega) = \frac{e^2/m}{\omega_0^2 - \omega^2 - i\gamma\omega}, \tag{3.80}$$

leading to a complex refractive index:

$$n(\omega) = \left(1 + \frac{Ne^2}{m\varepsilon_0} \frac{\omega_0^2 - \omega^2 + i\gamma\omega}{(\omega_0^2 - \omega^2)^2 + \gamma^2\omega^2}\right)^{1/2} = n_R(\omega) + in_I(\omega). \tag{3.81}$$

The real part of the refractive index describes the usual optical phenomena of refraction and reflection. The meaning of the imaginary part becomes evident by writing the oscillating electric field explicitly as a function of distance z, using the refractive index instead of the propagation number k:

$$E(z) = E_0 \, e^{i\omega(n(\omega)z/c - t)} = E_0 \, e^{-n_I(\omega)\omega z/c} \, e^{i\omega(n_R(\omega)z/c - t)}. \tag{3.82}$$

The amplitude of the electric field now decreases exponentially with distance z. Since the intensity of light is given by the square of the amplitude, we write

$$I(z) = I(0)e^{-2n_I(\omega)\omega z/c} = I(0)e^{-a(\omega)z}. \tag{3.83}$$

Equation 3.83 has the usual form of Lambert's law, where the absorption coefficient $a(\omega)$ is defined as

$$a(\omega) = \frac{2n_I(\omega)\omega}{c}. \tag{3.84}$$

Hence, the imaginary part of the refractive index describes the loss of energy by light passing through matter due to absorption. A related quantity is the absorption cross-section $\sigma_a(\omega)$, which is defined as $\sigma_a(\omega) = a(\omega)/N$.

The phenomenon of Rayleigh scattering refers to scattering of light from very small particles, that is, from particles whose size is much smaller than the wavelength of light. Such particles can be approximated by an oscillating dipole, and the scattering phenomenon can be accurately described by considering the properties of radiation emanating from an electric dipole. Considering a dipole p oscillating along the x-axis, as shown in Figure 3.26, the electric field radiated by the dipole at a distance r is written as

$$E_{rad} = \frac{\sin\theta}{4\pi\varepsilon_0 c^2 r} \frac{d^2 p}{dt^2}. \tag{3.85}$$

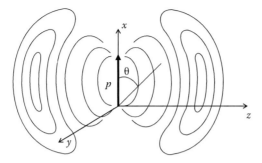

Figure 3.26 Geometry in light scattering.

The electric dipole of a small particle is assumed to be a Lorentz oscillating dipole driven by the electric field E_i propagating along z. From Equation 3.77, we have

$$p = \alpha(\omega)E_{0i}\cos(kz - \omega t). \tag{3.86}$$

Using this in Equation 3.85 gives

$$E_{\text{rad}} = \frac{\alpha(\omega)\omega^2 \sin\theta}{4\pi\varepsilon_0 c^2 r} E_{0i}\cos(kz - \omega t). \tag{3.87}$$

In terms of the radiated intensity of light, we may use the general relationship between the amplitude of an oscillating electric field E_0 and light intensity I given by the time-averaged Poynting vector as

$$I = |\langle \mathbf{S} \rangle_\tau| = \frac{c\varepsilon_0}{2} E_0^2, \tag{3.88}$$

to obtain the angle-dependent radiated intensity $I_{\text{rad}}(\theta)$,

$$I_{\text{rad}}(\theta) = \frac{\alpha^2(\omega)\omega^4 \sin^2\theta}{32\pi^2\varepsilon_0 c^3 r^2} E_{0i}^2 = \frac{\alpha^2(\omega)\omega^4 \sin^2\theta}{(4\pi\varepsilon_0)^2 c^4 r^2} I_i, \tag{3.89}$$

where I_i is the incident intensity along the z-axis. This equation gives the well-known fourth power dependence of scattered light with frequency. Light is scattered in a donut pattern with zero scattered intensity along the dipole axis x. The total scattered power Pwr at a distance r is obtained by integrating the angular coordinates across the whole space:

$$\text{Pwr} = r^2 \int_0^{2\pi} d\phi \int_0^{\pi} I_{\text{rad}}(\theta)\sin\theta\, d\theta = \frac{8}{3} r^2\pi I_{\text{rad}}(\theta) = \frac{8\pi}{3}\frac{\alpha^2(\omega)\omega^4}{(4\pi\varepsilon_0)^2 c^4} I_i. \tag{3.90}$$

The proportionality constant between Pwr and I_i has dimensions of area and is called the absorption cross section $\sigma_{sc}(\omega)$:

$$\sigma_{sc}(\omega) = \frac{8\pi\omega^4}{3c^4}\left(\frac{\alpha(\omega)}{4\pi\varepsilon_0}\right)^2.$$

(3.91)

Light propagating in matter in the z-direction is attenuated by scattering. It is possible to express the attenuation by an equation identical to Equation 3.83:

$$I(z) = I(0)e^{-N\sigma_{sc}(\omega)} = I(0)e^{-a_s(\omega)z}.$$

(3.92)

The quantity $a_s(\omega) = N\sigma_{sc}(\omega)$ is the scattering extinction coefficient.

3.2.2 Size-dependent scattering from dielectric spheres: Mie solutions

The phenomenon of Rayleigh scattering described in the previous section was an extreme case of scattering from particles with size $a \ll \lambda$, the wavelength of light. The general problem of scattering from particles of arbitrary size a was treated by G. Mie in 1908, who provided a rigorous solution to Maxwell's equations applied to homogeneous, isotropic, dielectric spheres interacting with plane electromagnetic waves. The results describe the scattering phenomenon for spherical particles of all sizes and refractive indices over all wavelength ranges. For nonspherical particles, the Mie theory is a good first-order approximation that correctly describes the qualitative effects of the change of size or refractive index. A strict formal solution is beyond the scope of this book; we summarize the main formulae and use numerical results of this theory to demonstrate the size-dependent effects in light scattering by nanoscale materials.

The Mie solutions of Maxwell's equations in spherical coordinates yield intensities of electric fields of scattered light observed at distances far greater than the wavelength of light (the so-called far-field zone) [21,22]. The solutions are in the form of spherical electric waves E_r and E_l corresponding to two polarizations, radial "r" and longitudinal "l". These waves are related to the plane waves of incident electric fields with polarizations E_{0r} and E_{0l}. The geometry of the problem is shown in Figure 3.27. And, E_r and E_l are expressed in the form

$$E_l = E_{0l}\frac{e^{-i(kr-\omega t)}}{ikr}S_2(\theta),$$

(3.93)

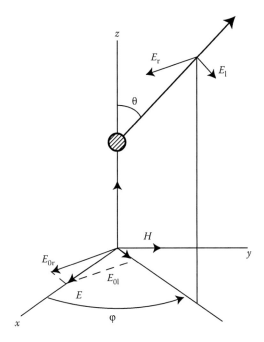

Figure 3.27 Definition of electric fields in Mie solutions of Maxwell's equations.

$$E_r = E_{0r} \frac{e^{-i(kr - \omega t)}}{ikr} S_1(\theta). \tag{3.94}$$

The functions $S_1(\theta)$ and $S_2(\theta)$ are amplitude scattering functions and given by

$$S_1(\theta) = \sum_{n=1}^{\infty} \frac{2n+1}{n(n+1)} \{a_n \pi_n(\cos\theta) + b_n \tau_n(\cos\theta)\}, \tag{3.95}$$

$$S_2(\theta) = \sum_{n=1}^{\infty} \frac{2n+1}{n(n+1)} \{b_n \pi_n(\cos\theta) + a_n \tau_n(\cos\theta)\}. \tag{3.96}$$

The angle-dependent functions $\pi_n(\cos\theta)$ and $\tau_n(\cos\theta)$ have been defined to simplify the notation. They contain Legendre polynomials of the first kind $P_n^1(\cos\theta)$ and are written as

$$\pi_n(\cos\theta) = \frac{P_n^1(\cos\theta)}{\sin\theta}, \tag{3.97}$$

$$\tau_n(\cos\theta) = \frac{dP_n^1(\cos\theta)}{d\theta}. \tag{3.98}$$

The parameters of the system are contained in the coefficients a_n and b_n. The size (radius) of a spherical particle a enters in the form of a size parameter, denoted by x, which relates the circumference of the sphere to the wavelength of light as $x = 2\pi a/\lambda$. The refractive index of the sphere n_1 relative to that of the medium n_2 is expressed as $m = n_1/n_2$. The problem of solving a particular scattering system reduces to the problem of calculating the coefficients a_n and b_n. These are expressed as

$$a_n = \frac{\psi_n(x)\psi_n'(mx) - m\psi_n(mx)\psi_n'(x)}{\xi_n(x)\psi_n'(mx) - m\psi_n(mx)\xi_n'(x)}, \tag{3.99}$$

$$b_n = \frac{m\psi_n(x)\psi_n'(mx) - \psi_n(mx)\psi_n'(x)}{m\xi_n(x)\psi_n'(mx) - \psi_n(mx)\xi_n'(x)}, \tag{3.100}$$

where ψ_n and ξ_n are Riccati–Bessel functions and the prime symbol indicates the first derivative with respect to position r.

The main results of the Mie solutions are best visualized in the form of numerical calculations. Figure 3.28 shows the angular (θ) dependence of the square of the scattered electric field intensity, which is equivalent to the intensity of light. The calculation is done for scattering of green light ($\lambda = 550$ nm) from spherical gold particles immersed in water. In the limit of a very small particle ($a = 5$ nm), the scattered light distribution corresponds to the electric dipole radiation (Rayleigh scattering). With increasing particle size, more light is scattered in the forward direction than in the opposite one and the distribution pattern becomes more complicated. This phenomenon is known as the Mie effect. For large particles ($a = 500$ nm), almost all light is scattered in the forward direction at $\theta = 0$.

Apart from the angular dependence of electric field intensity, other important results of the Mie solutions are the expressions for the cross-sections that describe the integrated spatial attenuation of the intensity of light upon interaction with the particle. The total attenuation is called extinction and is characterized by the extinction cross-section σ_{ex}. The attenuation is either by scattering described by σ_{sc} (see also Equation 3.91 for Rayleigh scattering) or by absorption described by σ_a (cf. Equation 3.84 for absorption by an oscillating dipole). The Mie theory gives

$$\sigma_{ex} = \frac{2\pi}{k^2} \sum_{n=1}^{\infty} (2n+1)\,\mathrm{Re}(a_n + b_n), \tag{3.101}$$

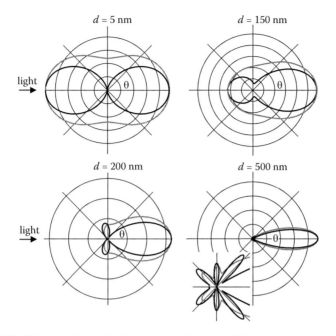

Figure 3.28 Calculated spatial distribution of scattered light intensity from a water suspension of spherical gold particles of different diameter *d* at the wavelength of 550 nm. Gray lines: natural light; black lines: linearly polarized light perpendicular to the plane of incidence. (The refractive index value of gold $n = 0.43 + i2.45$ was taken from P.B. Johnson and R.W. Christy, *Phys. Rev. B* 6, 1972, 4370.)

$$\sigma_{sc} = \frac{2\pi}{k^2} \sum_{n=1}^{\infty} (2n+1)(|a_n|^2 + |b_n|^2). \tag{3.102}$$

From the energy conservation law, it follows that

$$\sigma_a = \sigma_{ex} - \sigma_{sc}. \tag{3.103}$$

The related dimensionless efficiency factors are the cross-sections normalized by the geometrical cross-sections of the spheres. Thus,

$$Q_{ex} = \frac{\sigma_{ex}}{\pi a^2}, \quad Q_{sc} = \frac{\sigma_{sc}}{\pi a^2}, \quad \text{and} \quad Q_a = \frac{\sigma_a}{\pi a^2}. \tag{3.104}$$

The size dependence is readily seen in plots of the extinction, scattering, and absorption efficiency factors as functions of the particle size diameter *d* for metals with different optical constants (refractive indices): iron, gold, and silver (shown in Figure 3.29). For very small iron and gold particles, the Q_a factor is much larger than the Q_{sc} factor and a crossover of the two processes occurs at sizes between 80 and 120 nm. For silver particles that have a very small real part of the refractive index, the Q_{sc} dominates extinction from sizes of about 30 nm, and absorption is a relatively weak process. The size dependence of the efficiency factors for particles with different values of the imaginary part of the refractive index and with a constant real part is plotted in Figure 3.30. The figure clearly shows that with increasing imaginary refractive index, the crossover between absorption and scattering gradually shifts to larger particle sizes.

The Mie solutions also describe the variation of scattering and absorption efficiencies with the wavelength of light. Figure 3.31 shows calculated scattering efficiencies for transparent ($n_i = 0$) spherical particles of increasing size. With increasing size, the scattering efficiency shifts toward longer wavelengths and develops a characteristic oscillating pattern. For particles with nonzero imaginary refractive indices, such as gold

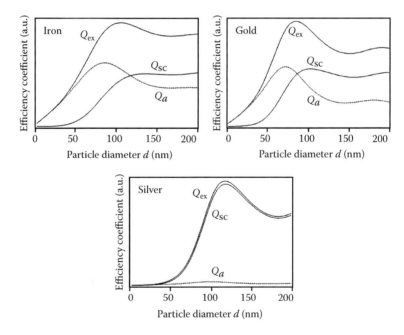

Figure 3.29 Calculated extinction, scattering, and absorption efficiency factors for water suspensions of three different metals at the wavelength of 550 nm. Iron: $n = 2.95 + i2.93$; gold: $n = 0.43 + i2.45$; silver: $n = 0.06 + i3.59$. (Refractive index data taken from P.B. Johnson and R.W. Christy, *Phys. Rev. B* 9, 1974, 5056; *Phys. Rev. B* 6, 1972, 4370.)

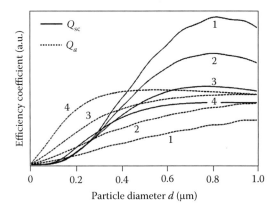

Figure 3.30 Calculated scattering and absorption efficiency factors as functions of the particle size d for a material with $n = 1.4 + in_1$ for increasing values of n_1. 1: $n_1 = 0.05$; 2: $n_1 = 0.1$; 3: $n_1 = 0.2$; 4: $n_1 = 0.4$. Wavelength is 550 nm.

particles dispersed in glass as shown in Figure 3.32, both scattering and absorption efficiencies exhibit a band centered between 500 and 600 nm, and for small particles, the absorption is stronger than scattering. With increasing particle size, the bands shift toward longer wavelengths, and the relative strengths of absorption and scattering are reversed for the largest particles.

3.2.3   Optical properties of metal nanoparticles: Plasmonics

One of the consequences of Mie solutions as seen from Figures 3.29, 3.30, and 3.32 is that for very small particles with large imaginary refractive

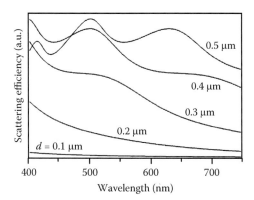

Figure 3.31 Calculated scattering efficiency as a function of the wavelength of light for transparent particles of the refractive index $n = 1.7$ and of different diameters d.

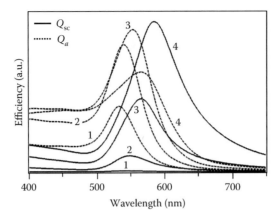

Figure 3.32 Calculated scattering (full lines) and absorption (dashed lines) efficiency as a function of the wavelength of light for spherical gold particles dispersed in glass of refractive index 1.5. Particle diameters: 1: $d = 20$ nm; 2: $d = 40$ nm; 3: $d = 60$ nm; 4: $d = 80$ nm. (Refractive index values of gold were taken from P.B. Johnson and R.W. Christy, *Phys. Rev. B* 6, 1972, 4370.)

indices, the extinction of light is dominated by absorption, while for increasing particle size the effect of scattering starts to dominate. This effect is well known for small particles of gold that are used to color stained glass. Very small gold particles predominantly absorb green light and appear red to the naked eye. In contrast, larger gold particles predominantly scatter green light and appear green. This effect follows from the limit of the Mie solutions for very small particles. For particles with sizes much smaller than the wavelength of light, the electric field across the particles is approximately constant and they are treated according to a so-called quasistatic approximation. The polarizability then scales with the particle radius a as

$$\alpha(\omega) = a^3 \, \frac{\varepsilon_1(\omega) - \varepsilon_2}{\varepsilon_1(\omega) + 2\varepsilon_2}, \tag{3.105}$$

where $\varepsilon_1(\omega)$ is the complex dielectric function of the particle and ε_2 the permittivity of the surrounding medium. We saw in Equation 3.91 that for vanishingly small particles the scattering cross-section is proportional to the square of the polarizability and thus to the sixth power of the diameter a, $\sigma_{sc} \propto |\alpha(\omega)|^2 \propto a^6$. It can also be shown that the absorption cross-section for such particles is linear with polarizability and thus scales with the third power of a, $\sigma_a \propto \text{Im}\{\alpha(\omega)\} \propto a^3$. These two different scaling laws explain the strong absorption of very small metal particles.

For very small spherical particles, one can also obtain approximate expressions for the frequency dependence of the extinction σ_{ex} and scattering σ_{sc} cross-sections that replace the exact Mie solution of Equations 3.101 and 3.102 [25]. It follows from the quasistatic approximation that

$$\sigma_{ex} = \frac{24\pi^2 a^3 \varepsilon_2^{3/2}}{\lambda} \frac{\varepsilon_1''(\omega)}{\left(\varepsilon_1'(\omega) + 2\varepsilon_2\right)^2 + \varepsilon_1''(\omega)^2}, \tag{3.106}$$

where $\varepsilon_1'(\omega)$ and $\varepsilon_1''(\omega)$ are the real and imaginary parts of the dielectric function of the particle and λ is the wavelength of light. Similarly,

$$\sigma_{sc} = \frac{16\pi^2 a^6 \varepsilon_2^{1/2}}{3\lambda} \frac{\left(\varepsilon_1'(\omega) - \varepsilon_2\right)^2 + \varepsilon_1''(\omega)^2}{\left(\varepsilon_1'(\omega) + 2\varepsilon_2\right)^2 + \varepsilon_1''(\omega)^2}. \tag{3.107}$$

It is now easy to see from Equations 3.106 and 3.107 that σ_{ex} and σ_{sc} are large when $\varepsilon_1'(\omega) = -2\varepsilon_2$ and at the same time $\varepsilon_1''(\omega)$ is small. These conditions correspond to the appearance of the absorption and scattering bands in the extinction spectra.

In addition to the size effects predicted by the Mie solutions of Maxwell's equations, there is another size effect in very small metallic particles that can only be explained in terms of a microscopic model of metals. Metals are characterized by partly filled valence or conduction bands. Electrons at the top of the energy distribution in such bands behave essentially as free electrons moving within the metal. Interaction of such free electrons in macroscopic-sized metallic samples with the electric field of light produces large oscillations of charge near the surface of the sample. Such oscillations are called surface plasmons (or surface plasmon polaritons). Further, these charge density oscillations give rise to a strong optical near-field in the vicinity of the interface. These surface plasmon phenomena can be described by a simple Drude model of metals, which is a modified damped Lorentz oscillator model of Equation 3.79, where the spring constant is set to zero. The damping γ now corresponds to the resistance of the metal and is given by the mean free path of the electrons between collisions with lattice atoms. The equation

$$\frac{d^2 x}{dt^2} + \gamma \frac{dx}{dt} = \frac{e}{m} E_0 e^{-i(\omega t - kz)} \tag{3.108}$$

now has a solution in terms of dielectric function as

$$\varepsilon_f(\omega) = 1 - \frac{\omega_{pf}^2}{\omega^2 - i\gamma\omega}, \tag{3.109}$$

and when separated into real and imaginary parts as

$$\varepsilon_f(\omega) = 1 - \frac{\omega_{pf}^2}{\omega^2 + \gamma^2} + i\frac{\gamma\omega_{pf}^2}{\omega(\omega^2 + \gamma^2)}. \tag{3.110}$$

The term $\omega_{pf} = \sqrt{N_f e^2/(m\varepsilon_0)}$ is the plasma frequency and N_f the volume density of free electrons. The free electrons in the metal are responsible for the large imaginary part of the refractive index and high reflectivity of the metal. The Drude model is accurate for the analysis of the properties of simple metals such as aluminum, but fails to describe the optical response of certain noble metals. For example, to account for the optical properties of silver, it is necessary to consider a substantial contribution of bound electrons, that is, electrons that behave like classical Lorentz oscillators. Physically, such electrons correspond to lower-lying d shells of metal atoms, and the optical transitions are called interband transitions. Equation 3.79 is solved in terms of the dielectric function as

$$\varepsilon_b(\omega) = 1 + \frac{\omega_{pb}^2}{\omega_0^2 - \omega^2 - i\gamma\omega}, \tag{3.111}$$

or separated into real and imaginary parts as

$$\varepsilon_b(\omega) = 1 + \frac{\omega_{pb}^2(\omega_0^2 - \omega^2)}{(\omega_0^2 - \omega^2)^2 + \gamma^2\omega^2} + i\frac{\gamma\omega_{pb}^2\omega}{(\omega_0^2 - \omega^2)^2 + \gamma^2\omega^2}. \tag{3.112}$$

The plasma frequency ω_b now contains the volume density of bound electrons, N_b. The response of noble metals is described by a composite dielectric function

$$\varepsilon(\omega) = \varepsilon_f(\omega) + \varepsilon_b(\omega). \tag{3.113}$$

Reduction in the size from bulk to nanometer scales leads to several important modifications. When a nanometer-sized metal particle interacts with the electric field of light, an overall displacement of the electron cloud with respect to the nuclei lattice is produced, as shown in Figure 3.33. The attractive restoring Coulomb force of the lattice then results in oscillations characterized by a specific resonant frequency. To distinguish this phenomenon from bulk surface plasmons, it is referred to as a localized plasmon. With decreasing particle size, the overall displacement becomes smaller and the resonant frequency increases. The resonant frequency gives rise to narrow absorption and scattering efficiency bands in the visible range of the spectra of metal nanoparticles, and the process is the physical interpretation of the theoretical Mie solutions.

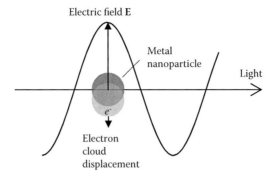

Electric field E

Metal nanoparticle

Light

e⁻

Electron cloud displacement

Figure 3.33 Schematic description of localized plasmon oscillation of a spherical metallic nanoparticle.

A different size effect is observed when the dimensions of particles are reduced to below the mean free path of electrons. The damping constant γ, which for bulk materials is given by the inverse of the time between collisions of free electrons, increases in nanoparticles because of additional collisions with the surface of the sample. The effective damping γ_{eff} in the particles is written as

$$\gamma_{eff} = \gamma + \frac{v_F}{L},\qquad(3.114)$$

where v_F is the electron velocity and L is the mean free path for collisions with the boundary, which is related to the particle radius as $L = 4a/3$. Further, to a good approximation, for many metals, $\omega_{pf} \gg \gamma$ and the free-electron bulk dielectric function (Equation 3.108) is given as

$$\varepsilon_f(\omega) = 1 - \frac{\omega_{pf}^2}{\omega^2} + i\,\frac{\gamma\omega_{pf}^2}{\omega^3}.\qquad(3.115)$$

Taking into account Equation 3.114, the size effect due to the reduced mean free path in nanoparticles is expressed in the imaginary part of the dielectric function as

$$\varepsilon_f''(\omega) = \frac{\gamma_{eff}\omega_{pf}^2}{\omega^3} = \frac{\omega_{pf}^2}{\omega^3}\left(\gamma + \frac{3v_F}{4a}\right).\qquad(3.116)$$

Figure 3.34 illustrates the various size effects that give rise to the optical properties of colloidal gold particles. As shown in the calculated absorption spectra of gold nanoparticles in Figure 3.32, the shift and broadening of the absorption peak for particles larger than about 10 nm

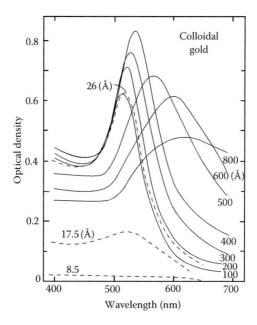

Figure 3.34 Absorption spectra of colloidal gold nanoparticles with different sizes. Full lines: size effects described by Mie solutions; dashed lines: size effects due to reduced mean free path of electrons. (Reproduced from Bohren, C.F. and Huffman, D.R., *Absorption and Scattering of Light by Small Particles*, Wiley-Interscience, New York, 1983. With permission.)

can be explained by the Mie theory using bulk values of the refractive index. Changes of the shape, intensity, and position of bands for small particles (below ~2.5 nm) are due to the effects of the reduced mean free paths of electrons.

The above treatment is valid for ideal spherical particles. For particles of nonspherical shape, the optical properties are more complicated. For example, in rod-shaped gold particles, the absorption band splits into two—one corresponding to electron oscillations along the rod (longitudinal) and the other to oscillations perpendicular to the rod long axis (transversal). The effect is shown in Figure 3.35. The underlying physics of the metallic particle–light interaction can be understood by analysis of extinction spectra of a pair of interacting gold particles. In Figure 3.36, the spectra are plotted for different distances between a pair of particles, ranging from 450 to 150 nm. With decreasing interparticle distance, the spectrum shifts to lower frequencies (higher wavelengths) and broadens. The physical interpretation of the plasmon resonance shift assumes a single dipole oscillator in isolated particles that respond to perturbation by light. The oscillator is schematically shown in Figure 3.36 as a spring

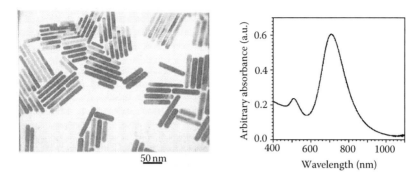

Figure 3.35 Left: Transmission electron microscope image of Au nanorods with a mean aspect ratio of 7.6. Right: absorption spectra of suspended gold nanorods solution with a mean aspect ratio of 5.2 showing splitting of the absorption into transversal band at 520 nm and longitudinal band at 720 nm. (Reprinted from Y.Y. Yu et al., *J. Phys. Chem. B*, 101, 1997, 6661. With permission.)

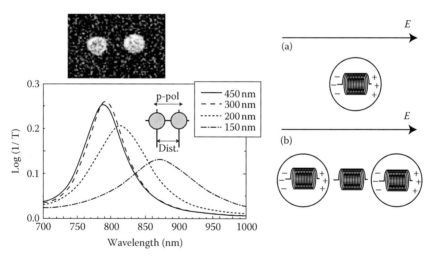

Figure 3.36 Top: Scanning electron microscope image of a pair of gold particles of 150 nm diameter prepared by electron-beam lithography. Bottom left: Extinction spectra of gold particle pairs for different interparticle distances obtained with light polarized along the pair common axis. Bottom right: Schematic interpretation of the plasmon resonance based on single dipole oscillator (a) and a pair of coupled dipole oscillators (b). (Reprinted from W. Rechberger et al., *Opt. Commun.* 220, 2003, 137. With permission.)

connecting the displaced electron cloud and the positively charged matrix region. When two particles are brought together so that their oscillators can interact, coupled oscillations occur—as shown in Figure 3.36 by the three connected springs—and the frequency of the resulting oscillations is smaller due to the increased effective length of the combined springs. In the extinction spectra, this effect appears as a red shift of the spectral peak that starts at a particle distance of 300 nm and is prominent for 150 nm.

3.2.4 Local field enhancement and surface-enhanced Raman scattering

The effect of electron cloud displacement in metallic nanoparticles upon their interaction with the electric field of light, as shown schematically in Figure 3.33, gives rise to absorption bands in the visible spectrum of light. At the same time, the periodic charge displacement results in an oscillating electric dipole, which in turn is a source of electromagnetic radiation. At distances far from the particle (far field), this radiated field is observed as scattering. At or near the particle surface, the oscillating dipole also results in strong enhancement of the electric field (near field). This local plasmon enhancement is due to charge present at the surface, and is a nanoparticle analogy of the near-field created by surface plasmons in bulk metallic films. In the quasistatic approximation, the vector of an electric field $\mathbf{E}_{LP}$ at a radial distance r from a metal nanoparticle surface can be expressed as [26]

$$\mathbf{E}_{LP} = E_0\hat{\mathbf{x}} - \alpha(\omega)E_0\left[\frac{\hat{\mathbf{x}}}{r^3} - \frac{3x}{r^5}(x\hat{\mathbf{x}} + y\hat{\mathbf{y}} + z\hat{\mathbf{z}})\right]. \tag{3.117}$$

Here, E_0 is the magnitude of the incident electric field polarized in the x-direction, $\alpha(\omega)$ is the particle polarizability given by Equation 3.105, and $\hat{\mathbf{x}}$, $\hat{\mathbf{y}}$, $\hat{\mathbf{z}}$ are unit vectors. The second term in Equation 3.117 describes the spatial distribution of electric fields of induced electric dipoles of magnitude, $\alpha(\omega)E_0$. In terms of the particle radius a, Equation 3.115 is rewritten using Equation 3.105 as

$$E_{LP} = \frac{\varepsilon_1(\omega) - \varepsilon_2}{\varepsilon_1(\omega) + 2\varepsilon_2}\left(\frac{a}{a+r}\right)^3 E_0. \tag{3.118}$$

The field E_{LP} in Equation 3.118 is taken along the oscillation direction of E_0. The field enhancement $A(\omega)$ is defined as [29]

$$A(\omega) = \frac{E_{LP} + E_0}{E_0} = \frac{\varepsilon_1(\omega) - \varepsilon_2}{\varepsilon_1(\omega) + 2\varepsilon_2}\left(\frac{a}{a+r}\right)^3 + 1. \tag{3.119}$$

The field enhancement can be particularly strong when the real part of the particle dielectric function $\varepsilon_1'(\omega)$ is equal to $-2\varepsilon_2$ (the condition for resonant frequency), and at the same time, the imaginary part of $\varepsilon_1''(\omega)$ is small. The field enhancement in the vicinity of a silver nanoparticle is shown in Figure 3.37. Note that the near-field is enhanced most in the direction of the applied electric field because it is the result of quasistatic charge induced on the surface. In contrast, the radiated electric field (far field) is strongest in directions perpendicular to the incident electric field.

Strong local field enhancement near the surface of metallic nanostructures can be exploited for a range of applications. For example, it can be used in aperture-less scanning near-field microscopy to excite fluorescence from molecules located close to a sharp metallic tip [30].

Local field enhancements that exist in nanostructured metallic surfaces also lead to strong enhancement of Raman scattering from molecules adsorbed on such surfaces. This phenomenon was first observed in the 1970s and is known as surface-enhanced Raman scattering (SERS). While it has been argued that chemical enhancement due to specific chemical interactions of the molecules with metallic surfaces is also partly responsible for SERS, the local field enhancement is by far the dominant effect in SERS. Local field enhancement is expressed using the enhancement factor $A(\omega_L)$ defined in Equation 3.119. The enhancement is effective in two ways. First, the incident laser light of intensity $I_0(\omega_L)$ is enhanced as $I_0(\omega_L)A(\omega_L)^2$, and induces Raman scattering, $P(\omega_S)$, from one molecule. Second, the Raman scattered light $P(\omega_S)$ is also enhanced by the same enhancement factor as $P(\omega_S)A(\omega_S)^2$ by passing through the

Figure 3.37 Local field enhancement calculated for a silver sphere with radius $a = 30$ nm in vacuum. Incident light propagates toward the figure and is polarized vertically. (Reproduced from L. Novotny and B. Hecht, *Principles of Nano-Optics*, Cambridge University Press, Cambridge, 2006. With permission.)

nanostructured sample. The resulting SERS signal can be expressed as [29]

$$P_{\text{SERS}}(\omega_S) = NI_0(\omega_L)|A(\omega_L)|^2|A(\omega_s)|^2\,\sigma_{\text{ads}}. \qquad (3.120)$$

Here, N is the number of adsorbed molecules and σ_{ads} is the Raman scattering cross-section for the adsorbed molecule. Assuming that the enhancement factors $A(\omega_L)$ for the laser and $A(\omega_S)$ for the Raman-scattered light are similar, we can write the total enhancement factor G_{SERS} for SERS relative to normal Raman scattering of free molecules in solution as

$$G_{\text{SERS}} = \frac{\sigma_{\text{ads}}}{\sigma_{\text{free}}}|A(\omega)|^4, \qquad (3.121)$$

where σ_{free} is the Raman cross-section of free molecules in solution. The ratio $\sigma_{\text{ads}}/\sigma_{\text{free}}$ reflects the effects of chemical enhancement in SERS. Normal Raman scattering is a very weak process, characterized by the σ_{free} cross-sections in the range 10^{-30}–10^{-25} cm^2. In comparison, the magnitude of the absorption cross-section σ_a for organic molecules is 10^{-17}–10^{-16} cm^2. However, the fourth power of the enhancement factor $A(\omega)$ in Equation 3.121 can result in giant SERS enhancement factors of up to 10^{14}, thus making the SERS process comparable in strength with absorption. It has been possible to measure SERS spectra of dye molecules adsorbed on silver nanoparticle clusters with very high sensitivity down to the level of single molecules. An example of such spectra for the organic dye Rhodamine 6G is shown in Figure 3.38.

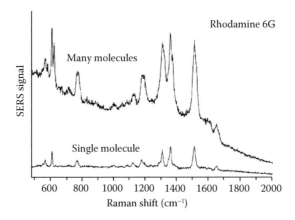

Figure 3.38 Many-molecule (top) and single-molecule (bottom) SERS spectra of Rhodamine 6G measured at different places on a fractal silver surface. (Reproduced from K. Kneipp and H. Kneipp, *Appl. Spectr.* 60, 2006, 322A. With permission.)

3.3 Size-dependent electromagnetic interactions: Particle–particle

This section describes interactions between nanoobjects that either involve electromagnetic fields or are initiated by interaction with light. These particle–particle interactions are strongly dependent on the distance between the particles and are mostly relevant on scales from 0.1 to tens of nanometers. The majority of this section will analyze interactions that result in transfer of electronic energy from one particle to another after absorption of light. Other types of interactions may result in the transfer of electrons between particles or due to strong coupling between the electron oscillators.

3.3.1 Radiative energy transfer

The simplest case of energy transfer is one in which a donor emits a photon, the energy traverses space as light, and is absorbed by an acceptor (Figure 3.39). The intensity of light decreases with distance from the source as r^{-2} and the radiative energy transfer efficiency follows the same donor–acceptor distance dependence. The probability that a photon emitted by a donor is absorbed by an acceptor is expressed as

$$p_{DA} \propto \frac{2.303[A]}{\phi_D} \int F_D(\omega)\varepsilon_A(\omega)d\omega, \qquad (3.122)$$

where [A] is the acceptor concentration, ϕ_D the donor fluorescence quantum efficiency in the absence of an acceptor, $F_D(\omega)$ the donor fluorescence, and $\varepsilon_A(\omega)$ the acceptor absorption spectra. At short distances (≤ 10 nm), the radiative contribution to energy transfer is negligible compared with the resonant mechanism, which is discussed in the following section. However, radiative transfer can be a dominant mechanism of energy transfer in dilute solutions, where it can influence fluorescence spectra and lifetimes.

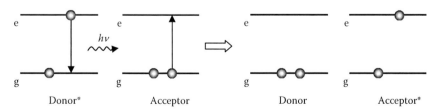

Figure 3.39 Scheme of radiative energy transfer between donor and acceptor molecules. Here, g and e are electronic ground and excited states.

3.3.2   Förster resonant energy transfer (FRET)

In Section 3.1.6, we discussed molecular dimers, which are weakly bound systems that interact strongly via dipole–dipole processes. The result is a coherent excitation delocalized over the system and a large shift of the energy levels of the system. Another important class of phenomena consists of systems that interact weakly by dipole–dipole interaction, so that excitation of an individual molecule in a pair of molecules can be considered as being independent. The weak interaction is due to relatively large intermolecular distances. Because resonant energy transfer occurs between two molecules or particles via dipole–dipole interaction, the only condition necessary for this process to occur is that the donors and acceptors possess strong transition dipole moments. Due to the spin selection rule for optical transitions, this condition usually implies energy transfer between donor and acceptor molecule singlet electronic states. However, there are cases where heavy-atom effects can partially lift the spin selection rule and triplet–singlet transitions (such as phosphorescence) can possess significant transition dipole moments. In such cases, the transfer of energy, for example, between donor triplet and acceptor singlet states can be an efficient process. The mechanism, which is effective on scales between 1 and 10 nm, is schematically depicted in Figure 3.40. Resonant energy transfer via dipole–dipole interaction is also referred to as Förster energy transfer after Th. Förster, who first derived the equation for the energy transfer rate. In deriving the equation, we begin with a general form of Fermi's golden rule for the transition rate between two quantum states:

$$\Gamma_{DA} = \frac{\pi}{2\hbar^2} |H_{DA}|^2 \, \delta(\omega_D - \omega_A). \tag{3.123}$$

Here, the subscripts D and A refer to the energy donor and energy acceptor molecules, respectively, ω_D and ω_A are the transition frequencies, Γ_{DA} is the energy transfer rate constant, and $|H_{AD}|$ is the matrix element of

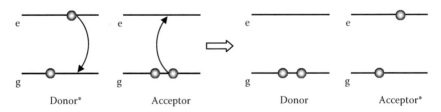

Figure 3.40 Scheme of resonant energy transfer between donor and acceptor molecules. Here, g and e are electronic ground and excited states.

the interaction Hamiltonian operator. We define the wave functions of the initial i and final f states of the pair of molecules as

$$\Psi_i = \psi_D^u \psi_A \quad \text{and} \quad \Psi_f = \psi_D \psi_A^u, \tag{3.124}$$

where $\psi_D^u \psi_A$ is a state with donors in the excited and acceptors in the ground state and $\psi_D \psi_A^u$ is a state with donors in the ground state and acceptors in the excited state. The matrix element of the interaction Hamiltonian operator can be then written as

$$|H_{DA}| = \int \Psi_f V_{DA} \Psi_i \, dr = \int \Psi_D \Psi_A^u V_{DA} \Psi_D^u \Psi_A \, dr. \tag{3.125}$$

The interaction energy operator V_{DA} is due to dipole–dipole interaction between transition dipoles μ_D and μ_A or, in other words, due to the potential energy of dipole μ_A in the electric field of dipole μ_D. Note that we encountered this energy in Equation 3.55, which we can rewrite using the notation of this section as

$$V_{DA} = \frac{1}{4\pi\varepsilon_0} \left(\frac{\mu_A \cdot \mu_D}{r^3} - \frac{3(\mu_A \cdot r)(\mu_D \cdot r)}{r^5} \right), \tag{3.126}$$

where ε_0 is the permittivity of vacuum. It is customary to define the relative position and distance of the two transition dipoles using the angles θ_D, θ_A, and θ_T and the distance r, as defined in Figure 3.41. The interaction energy V_{DA} is then written using scalar notation as

$$V_{DA} \propto \frac{\kappa \mu_D \mu_A}{r^3}, \tag{3.127}$$

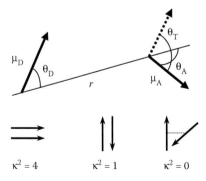

Figure 3.41 Definition of the symbols for FRET orientation factor and typical values of the factor for specific dipole orientations.

where the quantity κ is called the orientational factor and expressed as

$$\kappa = \cos\theta_T - 3\cos\theta_D\cos\theta_A. \tag{3.128}$$

The factor κ describes the contribution of the relative orientation of the donor and acceptor molecules to the interaction energy, and it can assume values between -2 and 2, as also shown in Figure 3.41. Substituting Equation 3.127 into Equation 3.125, we write the interaction Hamiltonian operator as

$$|H_{DA}| \propto \frac{\kappa}{r^3}\int \psi_D\psi_A^u(\hat{\mu}_D\hat{\mu}_A)\psi_D^u\psi_A\,dr. \tag{3.129}$$

Here, the symbols $\hat{\mu}_D$ and $\hat{\mu}_A$ now represent the transition dipole moment operators. The energy transfer rate is

$$\Gamma_{DA} \propto \frac{\kappa^2}{r^6}\left|\int \psi_D\psi_A^u(\hat{\mu}_D\hat{\mu}_A)\psi_D^u\psi_A\,dr\right|^2 \delta(\omega_D-\omega_A). \tag{3.130}$$

Separation according to integration coordinates of the D and A molecules leads to

$$\Gamma_{DA} \propto \frac{\kappa^2}{r^6}\left|\int \psi_D\hat{\mu}_D\psi_D^u\,dr_D\right|^2 \left|\int \psi_A^u\hat{\mu}_A\psi_A\,dr_A\right|^2 \delta(\omega_D-\omega_A). \tag{3.131}$$

Denoting the matrix element of the transition dipole moments of the donor and acceptor molecules as

$$|m_D| = \left|\int \psi_D\hat{\mu}_D\psi_D^u\,dr_D\right| \quad \text{and} \quad |m_A| = \left|\int \psi_A^u\hat{\mu}_A\psi_A\,dr_A\right|, \tag{3.132}$$

and using the following property of the delta function

$$\delta(\omega_D-\omega_A) = \int_0^\infty \delta(\omega_D-\omega)\delta(\omega-\omega_A)\,d\omega, \tag{3.133}$$

we rewrite Equation 3.131 for the transfer rate as

$$\Gamma_{DA} \propto \frac{\kappa^2}{r^6}\int_0^\infty |\mu_D|^2\,\delta(\omega_D-\omega)|\mu_A|^2\,\delta(\omega-\omega_A)\,d\omega. \tag{3.134}$$

Equation 3.134 already shows the well-known dependence of the transfer rate on the sixth power of the distance r. We can further manipulate this equation using the following relationships between the quantum-mechanical quantity of transition dipole moment and the macroscopic observables:

$$|\mu_A|^2 \delta(\omega - \omega_A) \propto \frac{\varepsilon_A(\omega)}{\omega}, \tag{3.135}$$

where $\varepsilon_A(\omega)$ is the acceptor absorption spectrum, and

$$|\mu_D^2| \frac{\pi}{3\varepsilon_0 \hbar^2} = B_{12} = \frac{\pi^2 c^3}{\hbar \omega^3} A_{12}, \tag{3.136}$$

where B_{12} and A_{12} are Einstein's coefficients for stimulated and spontaneous emission. Further, Einstein's coefficient A_{12} is related via the radiative lifetime τ_R to the fluorescence quantum efficiency ϕ_F and fluorescence lifetime τ_F as

$$A_{12} = \frac{1}{\tau_R} = \frac{\phi_F}{\tau_F}. \tag{3.137}$$

Combining the fluorescence quantum efficiency ϕ_F with the lineshape approximated by the delta function $\delta(\omega_D - \omega)$ leads to the fluorescence spectrum

$$\phi_F \delta(\omega_D - \omega) = F_D(\omega). \tag{3.138}$$

Substituting Equations 3.135 through 3.138 into Equation 3.134, we obtain for the energy transfer rate

$$\Gamma_{DA} \propto \frac{\kappa^2}{r^6 \tau_F} \int_0^\infty \varepsilon_A(\omega) F_D(\omega) \omega^{-4} \, d\omega. \tag{3.139}$$

It is further customary to use a normalized fluorescence spectrum $\bar{F}_D(\omega) = F_D(\omega)/\phi_F$ whereby Equation 3.139 changes to

$$\Gamma_{DA} \propto \frac{\kappa^2 \phi_F}{r^6 \tau_F} \int_0^\infty \varepsilon_A(\omega) \bar{F}_D(\omega) \omega^{-4} \, d\omega. \tag{3.140}$$

Equation 3.140 now shows the basic physics of the Förster energy transfer. The transfer rate is proportional to the overlap of the normalized donor fluorescence spectrum and the acceptor absorption spectrum; it decreases with sixth power of the donor–acceptor distance and depends on the donor fluorescence quantum efficiency and lifetime and the mutual orientation of the donor and acceptor transition dipole moments. The complete expression for the energy transfer, expressed in the units of wavelength λ, is given by

$$\Gamma_{DA} = \frac{(9000 \ln 10)\kappa^2 \phi_F}{128\pi^5 n^4 N_A r^6 \tau_F} \int_0^\infty \varepsilon_A(\lambda)\bar{F}_D(\lambda)\lambda^4 \, d\lambda, \qquad (3.141)$$

where N_A is Avogadro's number, and the refractive index n appears because the interaction between the two dipoles occurs in a dielectric medium rather than in vacuum.

It is now possible to define the *Förster distance* R_0 as a distance between a donor and an acceptor at which the energy transfer rate is equal to the fluorescence rate of a donor in the absence of an acceptor, that is, at which $\Gamma_{DA} = 1/\tau_F$. It follows that

$$R_0^6 = \frac{(9000 \ln 10)\kappa^2 \phi_F}{128\pi^5 n^4 N_A} \int_0^\infty \varepsilon_A(\lambda)\bar{F}_D(\lambda)\lambda^4 \, d\lambda. \qquad (3.142)$$

For a given pair of donor and acceptor molecules, R_0 is a constant and the rate of energy transfer is simply given by

$$\Gamma_{DA} = \frac{1}{\tau_F}\left(\frac{R_0}{r}\right)^6. \qquad (3.143)$$

One way to experimentally determine the energy transfer is by measuring relative changes in the intensity of the absorption or fluorescence spectra of both donors and acceptors. An example of the spectral changes is shown in Figure 3.42, where energy transfer occurs from a donor, which is an α-naphthyl group at the carboxyl end of a polypeptide, to the energy acceptor, which is a dansyl group at the peptide imino end. The efficiency of the energy transfer decreases with increasing length of the peptide spacer. The increase in n from 1 to 12 corresponds to the increase in the donor–acceptor distance from 1.2 to 4.6 nm. In the case of zero energy transfer and when monitored at the acceptor emission wavelength, only the fluorescence excitation spectrum of the acceptor is observed. With increasing energy transfer efficiency, both the donor and acceptor can be detected. The effectiveness of the energy transfer is expressed in terms of transfer efficiency defined as

$$\phi_{DA} = \frac{\Gamma_{DA}}{1/\tau_F + \Gamma_{DA}} = \frac{1}{1 + (r/R_0)^6}, \qquad (3.144)$$

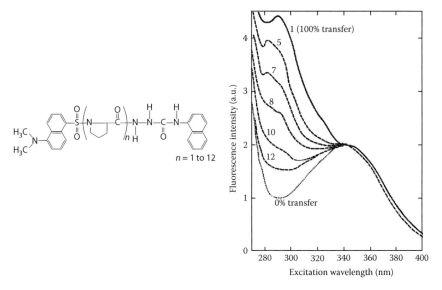

Figure 3.42 Left: Chemical structure of dansyl-(L-prolyl)$_n$-α-naphthyl. Donor, α-naphthyl group; acceptor, dansyl group. Right: Fluorescence excitation spectra measured for different lengths (n) of the peptide chain. (Reprinted from L. Stryer and R.P. Haugland, *Proc. Natl. Acad. Sci.* 58, 1967, 719. With permission.)

where again τ_F is the fluorescence lifetime of a donor in the absence of an acceptor. At R_0, the efficiency of the energy transfer is 0.5 and the energy is equally distributed between donor and acceptor fluorescence. Typically, R_0 is in the range of 2–5 nm. The distance dependence of the transfer efficiency is shown in Figure 3.43.

At this point, it is also interesting to compare the efficiencies of the radiative and Förster resonant (nonradiative) energy transfer as a function of the donor–acceptor distance. This is shown in Figure 3.44 in the form of energy transfer efficiency from polystyrene to tetraphenylbutadiene (TPB) as a function of the TPB acceptor concentration in solution. For longer distances, the radiative transfer is the dominant process and it has a maximum efficiency of about 15%. For shorter distances, there is a crossover below which nonradiative transfer starts to dominate.

Another way to study the energy transfer process is to measure the fluorescence lifetime of a donor in the presence of an acceptor and compare it with the donor fluorescence lifetime in the absence of an acceptor. With increasing efficiency of the energy transfer, the measured lifetime shortens compared with the case of zero energy transfer. An example is shown in Figure 3.45.

The Förster radius R_0 is comparable to the size of many proteins and to the thickness of biological membranes. It is thus possible to measure the efficiency of energy transfer between donor and acceptor labels in

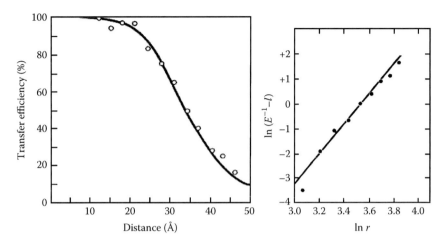

Figure 3.43 Left: Efficiency of energy transfer as a function of distance in dansyl-(L-prolyl)$_n$–α-naphthyl molecule for $n = 1$–12. Right: The same dependence in a log–log plot. The slope of the function is 5.9, in agreement with the r^{-6} dependence predicted by the Förster theory. (Reprinted from L. Stryer and R.P. Haugland, *Proc. Natl. Acad. Sci.* 58, 1967, 719. With permission.)

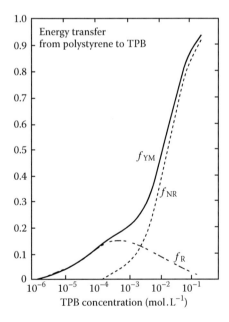

Figure 3.44 Efficiency of radiative f_R and nonradiative f_{NR} energy transfer from polystyrene to TPB in solution as a function of TPB concentration. (Reproduced from J.B. Birks, *Photophysics of Aromatic Molecules*, Wiley-Interscience, London, 1970. With permission.)

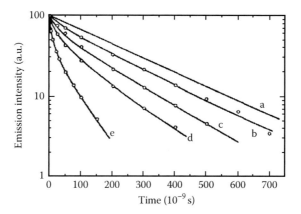

Figure 3.45 Fluorescence lifetime of a donor molecule of pyrene as a function of the concentration of an acceptor molecule, Sevron yellow. Acceptor concentrations: (a) 0; (b) 6.3×10^{-4} M; (c) 2.3×10^{-3} M; (d) 4.3×10^{-3} M; (e) 9.3×10^{-3} M. (Reprinted from R.G. Bennet, *J. Chem. Phys.*, 41, 1964, 3037. With permission.)

biological samples and to deduce from the results the distances between, for example, binding sites on proteins. Other examples are shown in Figure 3.46. Recently, Förster energy transfer is being increasingly used in the study of biochemical and biophysical processes, such as protein folding, at the single-molecule level.

3.3.3 Electron-exchange (Dexter) energy transfer

In donor–acceptor systems where the dipole–dipole energy transfer mechanism is negligible due to weak absorption or emission transition dipoles, the energy can be still transferred via electron-exchange mechanisms, first described by D.L. Dexter. Since the mechanism involves exchange of electrons between the donor and acceptor molecules, as schematically shown in Figure 3.47, it depends on the overlap of the respective molecular orbitals and is effective only on very short donor–acceptor distances (less than 1 nm). A theoretical description begins again with the transfer probability expressed by the general form of Fermi's golden rule

$$\Gamma_{DA} = \frac{\pi}{2\hbar^2} \left(\left| H_{DA}^e \right|^2 \right) \delta(\omega_D - \omega_A), \tag{3.145}$$

where $\left| H_{DA}^e \right|$ is the exchange interaction Hamiltonian operator. In the treatment of the Förster dipole–dipole energy transfer, this exchange term is neglected because it is much weaker than the dipole–dipole Coulomb term.

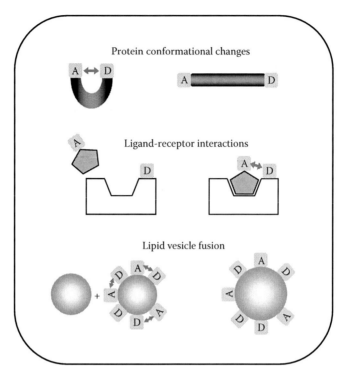

Figure 3.46 Examples of applications of the distance dependence of resonant energy transfer in the study of biological processes.

On the other hand, for weak transition dipole moments the Coulomb term is very small and the exchange term becomes the dominant interaction. The exchange term is expressed as

$$|H_{DA}^e| = \int \Psi_D(2)\Psi_A^u(1)(H_{DA}^e)\Psi_D^u(1)\Psi_A(2)\,dr, \qquad (3.146)$$

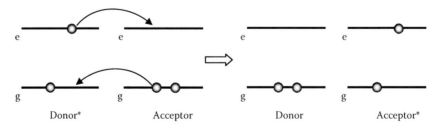

Figure 3.47 Scheme of electron-exchange energy transfer between donor and acceptor molecules. Here, g and e are electronic ground and excited states.

where the indices 1 and 2 refer to the different electrons originally located on the donor and acceptor molecules. It is now necessary to consider the spin state of the electrons by including spin wave functions χ as $\psi(1)\chi(1)$. The exchange term takes the form

$$|H^e_{DA}| = \int \psi_D(2)\chi_D(2)\psi^u_A(1)\chi^u_A(1)(H^e_{DA})\psi^u_D(1)\chi^u_D(1)\psi_A(2)\chi_A(2)\,dr\,d\sigma. \quad (3.147)$$

Since the Hamiltonian operator in Equation 3.147 does not operate on the spin wave functions, the integral can be separated as

$$|H^e_{DA}| = \int \psi_D(2)\psi^u_A(1)(H^e_{DA})\psi^u_D(1)\psi_A(2)\,dr \int \chi_D(2)\chi^u_A(1)\chi^u_D(1)\chi_A(2)\,d\sigma. \quad (3.148)$$

The second integral can be rewritten according to electron coordinates as

$$\int \chi_D(2)\chi^u_A(1)\chi^u_D(1)\chi_A(2)\,d\sigma = \int \chi_D(2)\chi_A(2)\,d\sigma_2 \int \chi^u_A(1)\chi^u_D(1)\,d\sigma_1. \quad (3.149)$$

As a result of the orthogonality of the spin wave functions of the same electrons, Equation 3.149 only has nonzero values when $\chi_D(2) = \chi_A(2)$ and $\chi^u_D(1) = \chi^u_A(1)$. The physical meaning of this condition is that energy transfer by the exchange mechanism can only occur when the ground states of the donor and acceptor molecules have the same spin, and at the same time, the excited states of donor and acceptor are in the same spin states. The condition for the ground states is largely satisfied since the ground states are usually singlet electronic states. The energy transfer can thus occur between donors and acceptors that have the same spin multiplicity of the excited states, that is, between singlet D and A or between triplet D and A. Singlet–singlet transitions are often dipole-allowed, and the energy transfer between excited singlet states is dominated by the dipole–dipole Förster mechanism. The Dexter exchange mechanism is thus often synonymous with triplet–triplet energy transfer. However, care must be taken in cases where S_0–T_1 transitions are partly allowed (efficient phosphorescence) or where S_0–S_1 transitions are forbidden.

The first integral of Equation 3.148 contains the exchange operator $H^e_{DA} = e^2/\kappa r$. The use of this operator leads to the following rate of energy transfer by electron-exchange mechanisms:

$$\Gamma^e_{DA} \propto \frac{2\pi}{\hbar}Z^2 \int_0^\infty \bar{\varepsilon}_A(\omega)\bar{F}_D(\omega)\,d\omega. \quad (3.150)$$

The integral in Equation 3.150 contains the overlap of normalized donor fluorescence and normalized acceptor absorption spectra. Since at least one of these transitions is usually spin-forbidden, the physical meaning of this integral is to express the condition for the energy difference between the donor and acceptor excited states. The quantity Z^2 is a function of the donor–acceptor distance r:

$$Z^2 \propto \exp\left(\frac{-2r}{L}\right), \qquad (3.151)$$

where L is the average Bohr radius for the excited and ground states of the donor and acceptor molecules, respectively. Since the energy transfer rate is an exponentially decreasing function of the distance r, energy transfer by the Dexter exchange mechanism occurs efficiently only over very short distances of less than ~1 nm.

3.3.4   Photo-induced electron transfer

In some aspects, the Dexter exchange mechanism of energy transfer is similar to another light-induced intermolecular interaction, electron transfer, which is usually treated as a subject of physical chemistry, and here only the main features of the process are discussed. The mechanism is schematically shown in Figure 3.48. Compared with the Dexter energy transfer, only the excited-state electron is transferred from the donor to the acceptor molecule. The result is the creation of D^+ and A^- ions. As with energy transfer, the condition necessary for electron transfer to occur is overlap of donor and acceptor molecular orbitals. The dependence of the transfer rate constant on the donor–acceptor distance is simplified as

$$\Gamma_{DA}^{el} \propto \exp[-\beta(r-L)], \qquad (3.152)$$

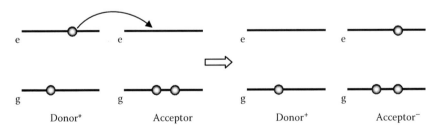

Figure 3.48 Scheme of photo-induced electron transfer between donor and acceptor molecules. Here, g and e are electronic ground and excited states.

where L is again the average Bohr radius and the coefficient β is a quantity inversely proportional to the overlap between the donor and acceptor electron orbitals. Equation 3.152 is an expression of the exponential decrease in electron transfer efficiency with increasing intermolecular separation. Besides the distance, other parameters affecting the electron transfer rate are the potential energy functions of the donor and acceptor, mutual orientation, shape and nodal character of the respective electron orbitals, and spin states.

3.4 Size-dependent interactions: Particle–light interactions in finite geometries

In the preceding sections, we treated size and distance-dependent effects that arise mainly due to the modification of the properties materials. A different type of size-dependent phenomena is observed when nanoscale-size matter interacts with light, which itself has modified properties. One such modification of light takes place in geometrically defined spaces where the light is confined. Such structures are called optical cavities and widely used in optical devices such as lasers. The cavities can be scaled down to scales of micrometers or hundreds of nanometers and are called microcavities. The first part of this section deals with interactions of light with materials placed inside such microcavities. For a rigorous treatment, it is necessary to invoke the quantum theory of light, but here we use the classical approach when possible and only state the results in other cases. The second part of this section describes interactions of material particles with light that are modified due to the presence of a dielectric interface within a distance of tens to hundreds of nanometers.

3.4.1 Optical interactions in microcavities

Microcavities are structures with highly reflecting surfaces where light can be confined in one, two, or three dimensions. In this respect, the confinement of light is reminiscent of the confinement of electrons in quantum wells, quantum wires, and quantum dots (Section 3.1). In free space, light propagates freely in any direction. The propagation is characterized by the wavevector, or propagation vector $\mathbf{k}$, the magnitude of which is given by $k = 2\pi/\lambda$, and by the wave equation for the electric field intensity of light $\mathbf{E}(\mathbf{r}, t)$,

$$\nabla^2 \mathbf{E}(\mathbf{r},t) = \frac{1}{c^2}\frac{\partial^2 \mathbf{E}(\mathbf{r},t)}{\partial t^2}. \tag{3.153}$$

Light confined in a microcavity can no longer propagate freely and this fact is expressed by a set of allowed values of the wavevector **k**. For confinement in three dimensions in a cubic cavity of length L, the allowed values of **k** are obtained from the wave equation (Equation 3.153), from Maxwell's equation for the electric field in vacuum:

$$\nabla \cdot \mathbf{E}(\mathbf{r}, t) = 0, \tag{3.154}$$

and from the boundary conditions. The solutions are written in the form of three separate equations for the electric field components:

$$
\begin{aligned}
E_x(\mathbf{r}, t) &= E_{0x} \exp(i\omega t)\cos(k_x x)\sin(k_y y)\sin(k_z z), \\
E_y(\mathbf{r}, t) &= E_{0y} \exp(i\omega t)\sin(k_x x)\cos(k_y y)\sin(k_z z), \\
E_z(\mathbf{r}, t) &= E_{0z} \exp(i\omega t)\sin(k_x x)\sin(k_y y)\cos(k_z z).
\end{aligned}
\tag{3.155}
$$

The components of the wavevector are now given by

$$k_x = \frac{\pi q_x}{L}, \quad k_y = \frac{\pi q_y}{L}, \quad \text{and} \quad k_z = \frac{\pi q_z}{L}, \tag{3.156}$$

where q_x, q_y, and q_z are mode numbers with values from 0, 1, 2 to n. Light can only exist inside the cavity in a finite number of defined modes. The situation is illustrated in Figure 3.49, which shows the allowed **k** values in a cubic microcavity of length L. The result of Equations 3.155 and

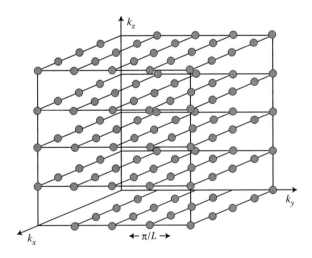

Figure 3.49 Allowed values of the wavevector **k** for light confined in a cubic microcavity.

3.156 expresses the well-known fact that light in a cavity can exist only if multiples of its half-wavelength match the cavity dimension. The number of allowed modes with wave vector magnitudes between k and $k + dk$ gives the DOS of light:

$$\rho_L(k)dk = \frac{k^2\,dk}{\pi^2},$$ (3.157)

or in terms of the angular frequency ω:

$$\rho_L(\omega)d\omega = \frac{\omega^2\,d\omega}{\pi^2 c^3}.$$ (3.158)

The DOS for 1D, 2D, and 3D light confinement in microcavities are schematically shown in Figure 3.50. Note the similarity of the light DOS with the electron DOS in quantum wells, quantum wires, and quantum dots, which were shown in Chapter 2. In particular, all of the DOS are concentrated into a few sharp spikes for light confined in all three dimensions.

When a material object is placed inside the microcavity, its interaction with light only occurs for light with allowed DOS. Depending on its strength, this interaction is classified into weak and strong. In the regime of weak interaction, the main effect is modification of the transition rate as described by Fermi's golden rule. A rigorous analysis using the quantum theory of light gives Fermi's golden rule that is similar to that of Equation 3.24, but which differs in that it describes the combined probability of transitions between two states of a molecule and two states of light. Keeping this in mind, we can use the semiclassical golden rule of Equation 3.25—which contains the density of the electronic states of a molecule—to describe transitions involving specific DOS of light. The molecule in such cases is considered to involve only two electronic

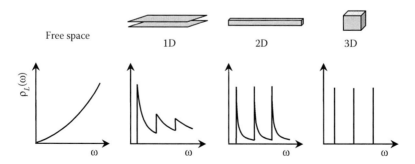

Figure 3.50 Schematic illustration of the DOS of light confined in one, two, and three dimensions.

levels. This simplified approach means using the DOS of Equation 3.158 in Equation 3.25:

$$\Gamma_{12} = \frac{\pi}{2\hbar^2} (E_0 \cdot \mu_{12})^2 \rho_L(\omega).$$ (3.159)

The densities of states described in Figure 3.50 are calculated for idealized microcavities with 100% reflective surfaces. In real cavities, the states are modified by the real properties of the system. As an example, Figure 3.51 shows light transmission through a planar Fabry–Perot cavity (1D light confinement case). For an ideal cavity, transmission occurs in sharp delta-function-like peaks. In real cavities, these peaks are broader and their width depends on the mirror reflectivity r via a parameter called finesse F:

$$F = \left(\frac{2r}{1 - r^2} \right)^2.$$ (3.160)

The cavity modes are defined by the 1D version of Equation 3.156:

$$k = \frac{\pi q}{d},$$ (3.161)

where d is the distance between the mirrors of the cavity.

In terms of light frequency, the modes that appear at v_q are described by

$$v_q = q \frac{c}{2d},$$ (3.162)

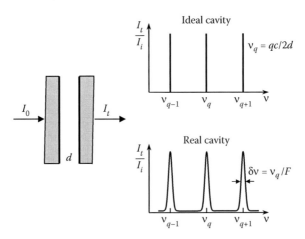

Figure 3.51 Transmittivity of an ideal (center) and real (right) Fabry–Perot cavity as depicted on the left.

and their width is given by

$$\delta\nu = \frac{\nu_q}{F}. \tag{3.163}$$

The mode's frequency position and width define the quality factor Q of the cavity as

$$Q = \frac{\nu_q}{\delta\nu}. \tag{3.164}$$

Variations of the transition rate due to modification of the DOS of light was first predicted as an increase in spontaneous emission rate (decrease in excited-state lifetime) by E.M. Purcell and is hence called the Purcell effect. In a 3D cavity with a quality factor Q and volume V, the lifetime is predicted to decrease by a factor F_P as

$$F_P = \frac{\tau_0}{\tau} = \frac{3}{4\pi^2}\left(\frac{\lambda_q}{n}\right)^3 \frac{Q}{V}, \tag{3.165}$$

where τ_0 and τ are the lifetimes outside and inside the cavity, respectively, and λ_q is the wavelength of the qth mode. The effect has been observed for InAs quantum dots placed in GaAs/AlAs pillars that function as Fabry–Perot cavities (Figure 3.52). Shortening of the lifetime

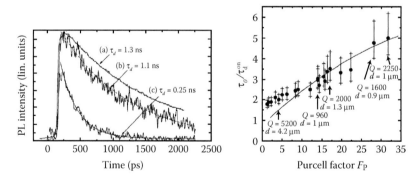

Figure 3.52 Left: Time-resolved photoluminescence intensity decays of InAs quantum dots in (a) bulk GaAs matrix, (b) inside the pillar out of resonance with the cavity mode, and (c) inside the pillar in resonance with the cavity mode. Right: Change of the decay time τ_d^{on} of the quantum dots on resonance with the cavity mode with the Purcell coefficient measured for different lengths d and quality factors Q of the cavity. (Reprinted from J.M. Gérard et al., *Phys. Rev. Lett.* 81, 1998, 1110. With permission.)

inside the cavity as well as different F_P factors for cavities of different Q-factors have been demonstrated. While light with higher densities of states compared with free space leads to an enhancement of spontaneous emission, the opposite, that is, inhibition of spontaneous emission for lower densities of states is also predicted and has been experimentally observed. Figure 3.53 shows changes of the lifetime for InGaAs quantum wells placed in a 3D microcavity formed by GaAs/AlAs pillars. The sides of the pillars are either left uncovered or are coated with a gold film for 3D confinement. Decreases in lifetime (emission enhancement) are observed when the emitted light wavelength is in resonance with the cavity modes for both cavities, while an increase in lifetime (emission inhibition) is observed for coated pillars when there is detuning from

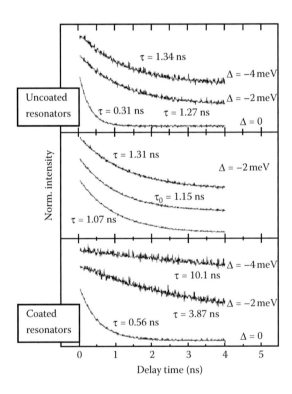

Figure 3.53 Top: Time-resolved photoluminescence intensity decays of InGaAs quantum wells placed inside uncoated and gold-coated GaAs/AlAs pillars. Bottom: Enhancement and inhibition of spontaneous emission as a function of detuning from the cavity mode. (Reprinted from M. Bayer et al., *Phys. Rev. Lett.* 86, 2001, 3168. With permission.)

the cavity mode wavelength. The combined effects of enhancement and inhibition are described by

$$\frac{\tau_0}{\tau} \cong \frac{2}{3} F_P \frac{\Delta\lambda_C^2}{\Delta\lambda_C^2 + 4(\lambda_C - \lambda_E)^2},$$ (3.166)

where λ_C is the cavity mode wavelength, λ_E is the quantum well emission wavelength, and $\Delta\lambda_C$ and $\Delta\lambda_E$ are their bandwidths.

The effect of spontaneous emission inhibition outside the cavity modes is also seen in fluorescence spectra of dye molecules inside a Fabry–Perot cavity, as shown in Figure 3.54. The fluorescence spectral shapes inside the cavity correspond well to the profile of the DOS given by the Q-factor. The Fabry–Perot cavity has one flat surface and one curved, as seen in the figure. Consequently, emission from single molecules occurs only at spatial locations where the emission wavelength matches the corresponding cavity mode. This fact gives rise to the concentric ring pattern in the microscopic image of single-molecule fluorescence as seen in Figure 3.54, illustrating well the cavity-length-dependent emission properties of dye molecules.

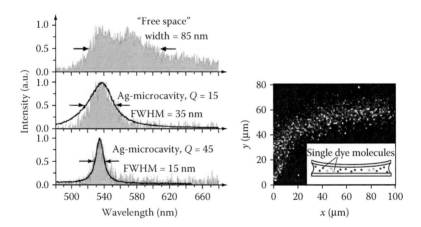

Figure 3.54 Left: Fluorescence spectra of single perylene dye molecules observed for molecules enclosed between the mirrors of a microcavity with Q-factors of 15 and 45 and in free space. The full lines are the respective transmission spectra of the cavity. Right: Fluorescence microscopic image of emission of individual perylene molecules located at the $q = 1$ mode of a Fabry–Perot cavity. (From M. Steiner et al., *Chem. Phys. Chem. 6*, 2005, 2190. With permission.)

3.4.2   Effects of dielectric interfaces

When a luminescent molecule or a nanoparticle is placed near metallic or dielectric surfaces (interface), its emission properties—fluorescence lifetime and quantum efficiency—can be significantly altered. This phenomenon is due to the interaction of the emitting dipole of the molecule with its own emitted electric field that is reflected from the surface. The interaction can result in either constructive or destructive interference, depending on the distance from the surface and polarization of the reflected electric field, which is determined by the orientation of the emitting dipole. Constructive interference leads to enhancement of spontaneous emission and shortening of the fluorescence lifetime, while destructive interference has the opposite effect. An example of this phenomenon is shown in Figure 3.55, where the fluorescence lifetime of Eu^{3+} ions in dibenzoylmethane complex is measured at different distances from a silver mirror [39]. The distance is controlled by placing the fluorescent molecule on fatty acid Langmuir–Blodgett films of varying thickness prepared on the silver mirror. The figure shows a periodic decrease and increase in the fluorescent lifetime with distance with the oscillation being damped at longer distances. The sudden drop in the lifetime at the shortest distances is due to a separate effect—energy transfer from the dye to the metallic layer. The changes in fluorescence lifetime can be conveniently explained by using the classical theory of spontaneous emission. In this theory, an emitting molecule in free space is approximated by an

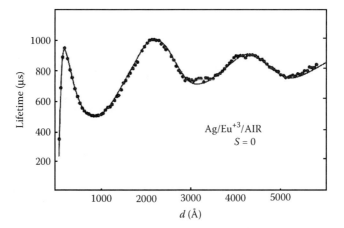

Figure 3.55 Fluorescence lifetime of europium dibenzoylmethane complex at different distances from a silver mirror. (Reprinted from R.R. Chance et al., *J. Chem. Phys.* 63, 1975, 1589; Experimental data from K.H. Drexhage, *J. Lumin.* 1,2, 1970, 693. With permission.)

oscillating nondriven electric dipole. Decay of the radiated intensity with time, which is equivalent to the loss of intensity due to spontaneous emission, is described by a damping constant β_0. The equation of motion for the dipole μ is expressed as

$$\frac{d^2\mu}{dt^2} + \beta_0 \frac{d\mu}{dt} + \omega_0^2 \mu = 0, \tag{3.167}$$

where ω_0 is the oscillator natural frequency. Equation 3.167 is equivalent to the left part of Equation 3.79 that we used in the classical description of the interaction of light with an atom. However, the physical meaning of the damping constant here is different. In Equation 3.79, the damping γ described absorption of the system, that is, damping of an oscillator driven by an external electric field. The damping constant β_0, on the other hand, describes a spontaneous attenuation of a self-oscillating dipole, the spontaneous emission.

When a molecule is placed near a metallic surface, its dipole interacts with the reflected electric field. Mathematically, this can be expressed by the equation of motion of a driven damped oscillator, where the driving force is the reflected field intensity E_R,

$$\frac{d^2\mu}{dt^2} + \beta_0 \frac{d\mu}{dt} + \omega_0^2 \mu = \frac{e^2}{m} E_R. \tag{3.168}$$

The interaction of the dipole with E_R will cause a shift in the resonant frequency and change in the emission rate. Using the trial solutions of Equation 3.168 in the form

$$\mu = \mu_0 e^{-i\omega t} e^{-\beta t/2}, \tag{3.169}$$

$$E_R = E_0 e^{-i\omega t} e^{-\beta t/2}, \tag{3.170}$$

we can obtain for the modified emission rate β the solution

$$\beta = \beta_0 + \frac{e^2}{2\mu_0 m \omega_0} \text{Im}\{E_0\}. \tag{3.171}$$

Assuming that in free space the emission rate is equal to the radiative rate, we can use the classical relationship for radiative rate:

$$\beta_0 = \frac{1}{4\pi\varepsilon_0} \frac{2e^2\omega_0^2}{3mc^3}, \tag{3.172}$$

and express the relative change of the emission rate near the surface as

$$\frac{\beta}{\beta_0} = 1 + \frac{3\pi\varepsilon_0 c^3}{\omega_0^3 \mu_0} \mathrm{Im}\{E_0\}. \tag{3.173}$$

The problem now translates to finding the reflected electric field from the planar metal or dielectrical surface. Instead of presenting a mathematically complex description of the reflected dipole field, the main effects are shown as results of simulations in Figure 3.56 [30]. The radiating dipole is in air at a distance h from either aluminum or glass surface. It can be oriented either parallel or perpendicular to the interface. The radiative lifetime (inverse of the emission rate) is plotted normalized with respect to that in free space. As can be expected, the effect is more pronounced for the metal surface due to increased reflectivity and for the parallel dipole orientation. Equation 3.173 contains only the reflected electric field as a parameter for the changed emission properties near the interface. An appropriate choice of the electric field can also describe the modified emission properties of molecules in other situations. For

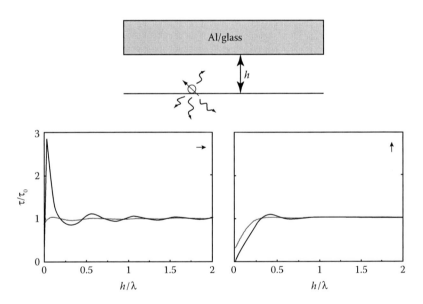

Figure 3.56 Top: Schematic description of the problem. Bottom: Normalized lifetime τ/τ_0 of a molecule located at a distance h from aluminum (black lines) or glass (gray lines) surfaces. The arrows indicate the dipole orientation with respect to a horizontal surface. (Reproduced from L. Novotny and B. Hecht, *Principles of Nano-Optics*, Cambridge University Press, Cambridge, 2006. With permission.)

example, the emission rate of a molecule changes in an inhomogeneous environment due to the interaction of the molecular dipole with its own electric field that is radiated and scattered back by the surrounding inhomogeneities. A proper expression for the scattered field E_0 in Equation 3.173 is used to describe the changes of emission lifetime and efficiency in such environment.

PROBLEMS

3.1 This problem illustrates the physical meaning of the transition dipole moment defined by Equation 3.17. Consider an optical transition between two energy levels of the molecule of anthracene. The molecular orbitals of the two levels (the highest occupied molecular orbital, and the lowest unoccupied molecular orbital) are shown in the figure, where the different shades indicate phases of the wave functions. Determine the orientation of the transition dipole moment with respect to the molecule for a transition between the two states.

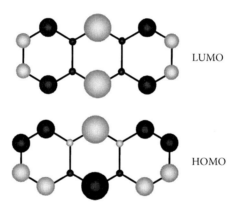

LUMO

HOMO

3.2 Calculate the Bohr radius of an exciton in the semiconductor CdSe. Assume the effective masses of electrons and holes to be $m_e^* = 0.13m_0$ and $m_h^* = 0.45m_0$, respectively, and the dielectric constant (permittivity) as being 9.3. Here, m_0 represents the electron rest mass, $m_0 = 9.1 \times 10^{-31}$ kg.

3.3 Using the radius obtained in Problem 3.2, calculate the wavelength of the lowest excitonic optical transition in bulk CdSe assuming a band gap energy of 1.89 eV. What would be the size of a CdSe nanocrystal required to produce an absorption peak at 520 nm?

3.4 The free-electron model assumes that delocalized π-electrons in linear conjugated molecules behave as free charges moving along a wire, and treats their wave functions and energies in terms of a particle-in-a-box. Calculate the wavelength of the lowest energy optical transition in the molecule octatetraene assuming that the transition takes place between the orbitals with principle quantum numbers 4 and 5. The length of the molecule can be assumed to be approximately 0.95 nm. How does the calculated result compare with the experimental value of 370 nm?

0.95 nm

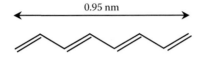

3.5 For systems that can be described in terms of a particle-in-a-box, derive the theoretical dependence of the transition dipole moment and the strength of optical transition on the size of the system, a.

3.6 Consider the absorption spectra of the naphthyl dimer series shown in Figure 3.15. Using the general form of the interaction energy given by Equation 3.58 in Equation 3.60, estimate the magnitude of the transition dipole moment of a naphthyl monomer, assuming that the average length of a carbon–carbon bond is 0.14 nm. Compare the result with the experimental value obtained from the extinction coefficient in Figure 3.14 using the approximate expression $|\mu|^2 \approx 9.2 \times 10^{-3} \int (\varepsilon/\nu)$ dν, where the transition dipole is given in units of Debye (1D = 3.33564×10^{-30} C m), ε is the molar extinction coefficient in M^{-1} cm^{-1} and ν is the frequency.

3.7 Consider scattering from very small particles (Rayleigh scattering). Show that for light of frequency of ω, which is much larger than any natural frequency ω_0, that the scattering cross-section can be described by the Thomson formula

$$\sigma_{SC}(\omega) \approx \frac{8\pi}{3} r_0^2,$$

where r_0 is the classical electron radius and given by $r_0 = e^2/4\pi\varepsilon_0\, mc^2$. Calculate the magnitude of r_0.

3.8 Calculate the Förster radius for dipole–dipole energy transfer between the organic dyes Rhodamine 101 (donor) and Rhodamine 700 (acceptor) dispersed randomly in an inert polymer matrix. The absorption and fluorescence spectra of each dye can be approximated by the sum of Gaussian functions, $\Sigma A_i \exp - ((\lambda - \lambda_i)/\delta\lambda_i)^2$, where λ_i is the Gaussian maximum and $\delta\lambda_i$ its width. The donor fluorescence and acceptor absorption spectra are characterized

Table 3.P1 Spectroscopic Parameters of the Dyes Rhodamine 101
and Rhodamine 700

	A_1	λ_1 (nm)	$\delta\lambda_1$ (nm)	A_2	λ_2 (nm)	$\delta\lambda_2$ (nm)
Rhodamine 101 fluorescence	0.0188	576	18	0.0061	607	37
Rhodamine 700 absorption	34600 $M^{-1}\,cm^{-1}$	595	30	94190 $M^{-1}\,cm^{-1}$	645	19

by the parameters shown in Table 3.P1. The polymer matrix has
a refractive index $n = 1.44$, and the fluorescence quantum efficiency of the donor in the absence of acceptor is 0.95.

3.9* The dependence of energy transfer rate on distance r as r^{-6} is
valid for energy transfer by dipole–dipole interaction between
a pair of isolated donor and acceptor molecules. Show that in
the case of energy transfer between monolayers of donor and
acceptor molecules separated by the distance d, the transfer rate
is proportional to d^{-4}. To derive the expression, refer to the simplified case in the figure and assume that each donor molecule
can transfer its energy to a large number of acceptor molecules
given by the molecular surface density σ_A. The expression will
be obtained by summing (integrating) the contributions to the
energy transfer rate from all acceptor molecules.

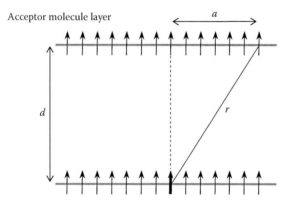

Acceptor molecule layer

Donor molecule layer

3.10 Consider the Purcell effect of enhancement and inhibition of
spontaneous emission in microcavities. Explain why it is possible to observe emission enhancement in a 1D cavity and
why for the observation of emission inhibition a 3D cavity is
necessary?

References

1. M. Schwoerer and H.C. Wolf, *Organic Molecular Solids*, Wiley-VCH, Weinheim, 2007.
2. R.S. Knox, Introduction to exciton physics. In: *Collective Excitations in Solids*, Plenum Press, New York, 1983.
3. P.N. Prasad, *Nanophotonics*, Wiley-Interscience, New Jersey, 2004.
4. D.S. Chemla, *Phys. Today* 46, 1993, 46.
5. M. Bruchez, M. Moronne, P. Gin, S. Weiss, and A.P. Alivisatos, *Science* 281, 1998, 2013.
6. G.D. Scholes and G. Rumbles, *Nat. Mater.* 5, 2006, 683.
7. J.B. Birks, *Photophysics of Aromatic Molecules*, Wiley-Interscience, London, 1970.
8. R.S. Becker, J.S. de Melo, A.L. Macanita, F. Elisei, *J. Phys. Chem.* 100, 1996, 18683.
9. J. Gierschner, H.-G. Mack, L. Lüer, and D. Oelkrug, *J. Chem. Phys.* 116, 2002, 8596.
10. E.G. Mc Rae, M. Kasha, *Physical Processes in Radiation Biology*, Academic Press, New York, 1964.
11. G.D. Scholes, K.P. Ghiggino, A.M. Oliver, and M.N. Paddon-Row, *J. Am. Chem. Soc.* 115, 1993, 4345.
12. A.A. Muenter, D.V. Brumbaugh, J. Apolito, L.A. Horn, F.C. Spano, and S. Mukamel, *J. Phys. Chem.* 96, 1992, 2783.
13. J. Hernando, J.P. Hoogenboom, E.M.H.P. van Dijk, J.J. García-López, M. Crego-Calama, D.N. Reinhoudt, N.F. van Hulst, and M.F. García-Parajó, *Phys. Rev. Lett.* 93, 2004, 236404.
14. H. Fidder and D.A. Wiersma, *Phys. Stat. Sol. B* 188, 1995, 285.
15. F.C. Spano and S. Mukamel, *Phys. Rev. A* 40, 1989, 5783.
16. S. Tatsuura, O. Wada, M. Tian, M. Furuki, Y. Sato, I. Iwasa, L.S. Pu, and H. Kawashima, *Appl. Phys. Lett.* 79, 2001, 2517.
17. B. Lounis, H.A. Bechtel, D. Gerion, P. Alivisatos, and W.E. Moerner, *Chem. Phys. Lett.* 329, 2000, 399.
18. M. Kuno, D.P. Fromm, H.F. Hamann, A. Gallagher, and D.J. Nesbitt, *J. Chem. Phys.* 112, 2000, 3117.
19. T. Huser, M. Yan, and L.J. Rothberg, *PNAS* 97, 2000, 11187.
20. C.W. Hollars, S.M. Lane, and T. Huser, *Chem. Phys. Lett.* 370, 2003, 393.
21. H.C. van de Hulst, *Light Scattering by Small Particles*, Dover Publications, New York, 1981.
22. C.F. Bohren and D.R. Huffman, *Absorption and Scattering of Light by Small Particles*, Wiley-Interscience, New York, 1983.
23. P.B. Johnson and R.W. Christy, *Phys. Rev. B* 9, 1974, 5056.
24. P.B. Johnson and R.W. Christy, *Phys. Rev. B* 6, 1972, 4370.
25. G.C. Papavassiliou, *Prog. Solid State Chem.* 12, 1979, 185.
26. K.L. Kelly, E. Coronado, L.L. Zhao, and G.C. Schatz, *J. Phys. Chem. B* 107, 2003, 668.
27. Y.Y. Yu, S.S. Chang, C.L. Lee, and C.R.C. Wang, *J. Phys. Chem. B* 101, 1997, 6661.
28. W. Rechberger, A. Hohenau, A. Leitner, J.R. Krenn, B. Lamprecht, and F.R. Aussenegg, *Opt. Commun.* 220, 2003, 137.
29. K. Kneipp, H. Kneipp, *Appl. Spectr.* 60, 2006, 322A.

30. L. Novotny and B. Hecht, *Principles of Nano-Optics*, Cambridge University Press, Cambridge, 2006.
31. L. Stryer and R.P. Haugland, *Proc. Natl. Acad. Sci.* 58, 1967, 719.
32. R.G. Bennet, *J. Chem. Phys.* 41, 1964, 3037.
33. B. Valeur, *Molecular Fluorescence*, Wiley-VCH, Weinheim, 2006.
34. R. Loudon, *The Quantum Theory of Light*, 3rd Ed., Oxford University Press, Oxford, 2000.
35. C. Weisbuch, H. Benisty, and R. Houdre, *J. Lumin.* 85, 2000, 271.
36. J.M. Gérard, B. Sermage, B. Gayral, B. Legrand, E. Costard, and V. Thierry-Mieg, *Phys. Rev. Lett.* 81, 1998, 1110.
37. M. Bayer, T.L. Reinecke, F. Weidner, A. Larionov, A. McDonald, and A. Forchel, *Phys. Rev. Lett.* 86, 2001, 3168.
38. M. Steiner, F. Schleifenbaum, C. Stupperich, A. Virgilio Failla, A. Hartschuh, and A.J. Meixner, *Chem. Phys. Chem.* 6, 2005, 2190.
39. K.H. Drexhage, *J. Lumin.* 1,2, 1970, 693.
40. R.R. Chance, A.H. Miller, A. Prock, and R. Silbey, *J. Chem. Phys.* 63, 1975, 1589.

chapter four

Magnetic and magnetotransport properties of nanoscale materials

4.1 Fundamentals of magnetism

We start this chapter with a brief review of some of the basic physical phenomena required for understanding magnetism. First, we introduce some examples of magnetic ordering, including ferromagnetism, antiferromagnetism, and spin glass. The basis of the magnetic properties of materials is mainly a result of interactions between spins of constituent atoms in the material and the most important interaction in magnetic materials is exchange interaction, which originates from the Pauli exclusion principle. The nature of the exchange interaction determines the type of magnetic ordering exhibited by materials. In order to observe how magnetic ordering occurs, we discuss the mean field theory for ferromagnetism by derivation of the Curie–Weiss law. The mean field theory can easily be applied for analysis of other magnetic ordering, and the approach there is a simple extension of that of ferromagnetism. Thus, from this viewpoint, discussing the mean field theory of ferromagnetism is instructive.

4.1.1 Magnetic ions and magnetic ordering

Electrons in an atom are known to occupy electronic orbitals so as to obey the Pauli exclusion principle, and the ground state is determined mainly by Hund's rule. Now, Hund's rule tells us that, first, the ground state of an atom is such that the total spin angular momentum $\mathbf{S} = \mathbf{s}_1 + \mathbf{s}_2 + \cdots$ is the sum of electron spins, $\mathbf{s}_i$, and must be maximized; second, the total orbital angular momentum $\mathbf{L} = \mathbf{l}_1 + \mathbf{l}_2 + \cdots$ is the sum of electronic orbital moments, $\mathbf{l}_i$, and must be the maximum. The angular momentum is directly related to the magnetic moment $\boldsymbol{\mu}$ of an atom as $\boldsymbol{\mu} = -\mu_B(2\mathbf{S} + \mathbf{L})$, where the magnetic moments of atoms in solids interact with each other to a certain degree. When the interaction between the magnetic moments of atoms is sufficiently small compared with the thermal energy $k_B T$, then all the magnetic moments freely rotate at a very high frequency in the GHz range. This magnetic state is called paramagnetic. However, when some interactions, which make the relative orientation of the magnetic moments parallel, are large enough to overcome the thermal energy, then all the magnetic moments align parallel to each other. This type of magnetism is called ferromagnetism. When the interaction forces the orientation of neighboring moments to be antiparallel, then no net magnetization remains, and we call this antiferromagnetism. When the magnetic moments of constituent ions are not identical, then a total magnetization remains nonvanishing even in the antiferromagnetic spin configuration, and such magnetic ordering is called ferrimagnetism. If magnetic moments are randomly frozen and fixed at a certain direction independent of time, the magnetism is called spin glass. These magnetic orderings and the magnetization processes are summarized in Figure 4.1.

4.1.2   *Exchange interaction*

Magnetism has its origin in the angular momentum (orbital angular momentum $\hbar l$ and spin angular momentum $\hbar s$) of electrons surrounding the nucleus of an atom as discussed before. According to quantum mechanics, the eigenvalue of l^2 is $l(l + 1)$ and that of l_z is quantized into

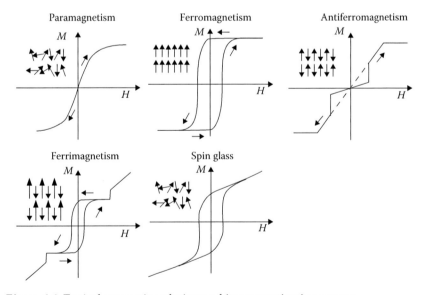

Figure 4.1 Typical magnetic ordering and its magnetization curves.

$(2l + 1)$ values, that is, l, $l - 1, \ldots -l$ ($l = 0, 1, 2, \ldots$), where we take the quantization axis to be along the z-direction. On the other hand, experiments show that the z component of the spin angular momentum, **s**, takes two values, so the value of **s** must be $1/2$, that is, the eigenvalue of $\mathbf{s}^2$ is $s(s + 1) = 3/4$ and that of $\mathbf{s}_z$ is either $1/2$ or $-1/2$. The eigenstates of a spin (spin function) are usually written as $\alpha(\xi)$ for $\mathbf{s}_z = 1/2$ (spin-up) and that for $\mathbf{s}_z = -1/2$, as $\beta(\xi)$ (spin-down), where ξ is the spin variable. Thus, let an electron occupy one of the single electron wave functions given by

$$\Psi(\tau) = \Psi_{i,\sigma}(\mathbf{r}, \xi) = \Psi(\mathbf{r})\chi_\sigma(\xi), \tag{4.1}$$

where $\Psi_i(\mathbf{r})$ is the orbital function, $\mathbf{r}$ is the position of the electron and $\chi_\sigma(\xi)$ is the spin function, that is, $\alpha(\xi)$ or $\beta(\xi)$ for $\sigma = 1$ and $\sigma = -1$, respectively. The spin functions, $\alpha(\xi)$ and $\beta(\xi)$, have the following elements for spin variable $\xi = +1$ and -1:

$$\begin{array}{ll} \alpha(+1) = 1 & \beta(+1) = 0 \\ \alpha(-1) = 0' & \beta(-1) = 1' \end{array} \tag{4.2}$$

or in the matrix representation

$$\alpha = \begin{pmatrix} 1 \\ 0 \end{pmatrix} \quad \text{and} \quad \beta = \begin{pmatrix} 0 \\ 1 \end{pmatrix}. \tag{4.3}$$

Also, the matrix representations of the x, y, and z components of spin **s** are given by

$$S_x = \begin{pmatrix} 0 & 1/2 \\ 1/2 & 0 \end{pmatrix}, \quad S_y = \begin{pmatrix} 0 & -i/2 \\ i/2 & 0 \end{pmatrix}, \quad \text{and} \quad S_z = \begin{pmatrix} 1/2 & 0 \\ 0 & -1/2 \end{pmatrix}. \tag{4.4}$$

Now, consider a two-electron system and let these electrons occupy the single electron states $\Psi_1(\tau_1)$ and $\Psi_2(\tau_2)$. Because the two-electron wave function $\Phi(\tau_1, \tau_2)$ must be antisymmetric with respect to the permutation of the coordinates of electrons 1 and 2, we choose a Slater determinant for the two-electron wave function, which to a first approximation is given by

$$\Phi(\tau_1, \tau_2) = \frac{1}{\sqrt{2}} \begin{vmatrix} \Psi_1(\tau_1) & \Psi_2(\tau_1) \\ \Psi_1(\tau_2) & \Psi_2(\tau_2) \end{vmatrix}. \tag{4.5}$$

If the two electrons occupy different orbitals, then the following four wave functions are possible for the system:

$$\Phi_1(\tau_1,\tau_2)=\frac{1}{\sqrt{2}}[\Psi_1(\mathbf{r}_1)\alpha(\xi_1)\Psi_2(\mathbf{r}_2)\beta(\xi_2)-\Psi_1(\mathbf{r}_2)\alpha(\xi_2)\Psi_2(\mathbf{r}_1)\beta(\xi_1)],$$

$$\Phi_2(\tau_1,\tau_2)=\frac{1}{\sqrt{2}}[\Psi_1(\mathbf{r}_1)\beta(\xi_1)\Psi_2(\mathbf{r}_2)\alpha(\xi_2)-\Psi_1(\mathbf{r}_2)\beta(\xi_2)\Psi_2(\mathbf{r}_1)\alpha(\xi_1)],$$

$$\Phi_3(\tau_1,\tau_2)=\frac{1}{\sqrt{2}}[\Psi_1(\mathbf{r}_1)\Psi_2(\mathbf{r}_2)-\Psi_1(\mathbf{r}_2)\Psi_2(\mathbf{r}_1)]\alpha(\xi_1)\alpha(\xi_2),$$

$$\Phi_4(\tau_1,\tau_2)=\frac{1}{\sqrt{2}}[\Psi_1(\mathbf{r}_1)\Psi_2(\mathbf{r}_2)-\Psi_1(\mathbf{r}_2)\Psi_2(\mathbf{r}_1)]\beta(\xi_1)\beta(\xi_2).$$

(4.6)

Using these wave functions as a basis of the system, the matrix elements of the Coulomb interaction e^2/r_{12} are calculated to be

$$\langle\Phi_i(\tau_1,\tau_2)|\frac{e^2}{r_{12}}|\Phi_j(\tau_1,\tau_2)\rangle=\begin{bmatrix} C & -J & 0 & 0 \\ -J & C & 0 & 0 \\ 0 & 0 & C-J & 0 \\ 0 & 0 & 0 & C-J \end{bmatrix},$$

(4.7)

$$C=\iint d\mathbf{r}_1\,d\mathbf{r}_2\Psi_1^*(\mathbf{r}_1)\Psi_2^*(\mathbf{r}_2)\frac{e^2}{r_{12}}\Psi_1(\mathbf{r}_1)\Psi_2(\mathbf{r}_2),$$

$$J=\iint d\mathbf{r}_1\,d\mathbf{r}_2\Psi_1^*(\mathbf{r}_1)\Psi_2^*(\mathbf{r}_2)\frac{e^2}{r_{12}}\Psi_2(\mathbf{r}_1)\Psi_1(\mathbf{r}_2),$$

(4.8)

where C and J are the Coulomb integral and the exchange integral, respectively.

Let us introduce another simple approach for calculating the exchange energy. Consider the following spin operator:

$$H_{exch}=-J(\tfrac{1}{2}+2\mathbf{s}_1\cdot\mathbf{s}_2)=-J(\tfrac{1}{2}+2s_1^z s_2^z)-J(s_1^+ s_2^- + s_2^+ s_1^-),$$

(4.9)

where the exchange integral is given as a parameter. When we calculate the matrix elements of this operator using the spin functions as a basis,

$$\begin{array}{l} \alpha(\xi_1)\beta(\xi_2) \\ \beta(\xi_1)\alpha(\xi_2) \\ \alpha(\xi_1)\alpha(\xi_2)' \\ \beta(\xi_1)\beta(\xi_2) \end{array}$$

(4.10)

the matrix elements are given as

$$
\begin{bmatrix}
0 & -J & 0 & 0 \\
-J & 0 & 0 & 0 \\
0 & 0 & -J & 0 \\
0 & 0 & 0 & -J
\end{bmatrix}.
\tag{4.11}
$$

We notice here that the matrix elements are the same as the exchange energy parts in Equation 4.7. Therefore, the spin operator of Equation 4.9 gives matrix elements equivalent to the exchange energy using the spin functions of Equation 4.10. This spin operator is called an exchange energy operator and enables us to calculate the exchange energy only using the spin operators and the spin functions, instead of direct calculations with the wave functions in Equation 4.6. It should be noted from Equation 4.9 that the relative spin orientation depends on the sign of the exchange integral J and that this arises mainly due to the Pauli exclusion principle.

4.1.3 Mean field theory of ferromagnetism

In the previous section, we discussed the exchange interaction between two electrons. A similar exchange interaction between the spins of atoms S_i and S_j is a simple extension of the aforementioned mathematical analysis, and the total exchange energy Hamiltonian operator of an N-spin system is written as

$$
H_{\text{exch}} = -2J \sum_{i<j} \mathbf{S}_i \cdot \mathbf{S}_j.
\tag{4.12}
$$

Assume that all the atoms are located in the same environment, and substituting the average $\langle \mathbf{S} \rangle$ for $\mathbf{S}_j$ in Equation 4.12 reduces this equation to

$$
H_{\text{exch}} = -2Jz\langle \mathbf{S} \rangle \cdot \mathbf{S}_i = \frac{2zJ}{N(g\mu_B^2)}(-Ng\mu_B\langle \mathbf{S} \rangle) \cdot g\mu_B \mathbf{S}_i = w\mathbf{M} \cdot g\mu_B \mathbf{S}_i,
$$

$$
w = \frac{2zJ}{N(g\mu_B^2)},
\tag{4.13}
$$

$$
\mathbf{M} = -Ng\mu_B\langle \mathbf{S} \rangle.
$$

Equation 4.13 tells us that the magnetic moment of $-g\mu_B\mathbf{S}_i$ is located in a magnetic field of $w\mathbf{M}$, which we call an effective magnetic field or Weiss field, originating from the exchange interaction between spin $\mathbf{S}_i$ and its surrounding spins.

To obtain an expression for the magnetization of an exchange-coupled spin system, we first introduce the Langevin–Debye theory, which gives the magnetization of a noninteracting system in an external magnetic field of **H**. Since the Zeeman energy of a magnetic moment $-g\mu_B\mathbf{S}$ is given by $g\mu_B S_z H$ in a magnetic field along the z-direction, the magnetization M is written as

$$M = \frac{N\sum_{S_z=-S}^{S} -g\mu_B S_z \exp(-g\mu_B S_z H/k_B T)}{\sum_{S_z=-S}^{S} \exp(-g\mu_B S_z H/k_B T)}. \tag{4.14}$$

Substituting with $x = g\mu_B S_z H/k_B T$, Equation 4.13 is reduced to

$$M = N \frac{g\mu_B S(d/dx)\sum_{S_z=-S}^{S} \exp(-xS_z/S)}{\sum_{S_z=-S}^{S} \exp(-xS_z/S)}$$

$$= Ng\mu_B S \frac{d}{dx}\log\left[\frac{\sinh(((2S+1)/2S)x)}{\sinh(x/2S)}\right] \tag{4.15}$$

$$= Ng\mu_B S\left[\frac{2S+1}{2S}\coth\left(\frac{2S+1}{2S}x\right) - \frac{1}{2S}\coth\left(\frac{x}{2S}\right)\right].$$

The $[\cdots]$ part in Equation 4.15 is called the Brillouin function and is termed $B_S(x)$. In the limiting case of $S \to \infty$, the Brillouin function is reduced to Langevin function of $\coth x - 1/x$. When the external magnetic field is sufficiently small, the magnetization is written as

$$M = Ng\mu_B S \frac{S+1}{3S} x = N(g\mu_B)^2 \frac{S(S+1)}{3k_B T} H. \tag{4.16}$$

Therefore, the magnetic susceptibility χ is obtained as

$$\chi = \frac{M}{H} = N(g\mu_B)^2 \frac{S(S+1)}{3k_B T} = \frac{C}{T},$$

$$C = N(g\mu_B)^2 \frac{S(S+1)}{3k_B}, \tag{4.17}$$

where the magnetic susceptibility is inversely proportional to temperature, and this is called Curie's law.

Since we have deduced an expression for magnetization in a noninteracting spin system located in a magnetic field H, we now calculate magnetization for a spin system with exchange interaction by substituting the corresponding effective magnetic field for the external magnetic field in Equation 4.15,

$$M = Ng\mu_B S B_S(x),$$

$$x = g\mu_B S \frac{2xJ}{N(g\mu_B)^2} \frac{(-Ng\mu_B\langle S \rangle)}{k_B T}. \tag{4.18}$$

If we take the sum of the Weiss field and the external magnetic field as H in Equation 4.16,

$$M = Ng\mu_B(S+1)\frac{x}{3}$$

$$= N(g\mu_B)^2 S(S+1)\frac{1}{3k_B T}(H + wM) \tag{4.19}$$

is obtained for a high-temperature limit ($x \to 0$). From Equation 4.19, the magnetization and magnetic susceptibility are written as

$$M = \frac{CH}{T - Cw},$$

$$\chi = \frac{C}{T - Cw}. \tag{4.20}$$

Equation 4.20 is called the Curie–Weiss law for a ferromagnetic spin system with an exchange interaction J. The argument in this section is based mainly on the assumption that the exchange interaction can be approximated as an effective magnetic field around the particular spin, and this sort of approach is called the mean field theory and is applicable to other spin systems.

4.2   Size and surface effects in 3D confined systems

The thermodynamic properties of nanostructures are greatly influenced by their size due to the large spacing of energy levels of quantized electronic states in the system. One of the most tremendous effects associated with system size is the Kubo effect on the thermodynamic properties of metallic nanoparticles, theoretically predicted in 1962 [1]. A pure statistical mechanical approach yields fascinating information about the physical properties of an ensemble of nanoparticles. In contrast, the ratio of surface to total volume is significantly larger in nanostructures, where surface-related effects appear in their physical properties. A typical example is magnetism of 4*d* transition metals, where ferromagnetism appears in thin films and nano-particles due to surface effects. In this section, we first discuss the theoretical aspects of these two effects in magnetic nanostructures and present some beautiful experimental demonstrations.

4.2.1 Quantization of electronic structures and the Kubo effect

We are familiar with the fact that hybridization of electron orbitals leads to the formation of continuous bands (as discussed in Chapter 1), whereas in nanostructures the properties of the individual electron orbitals are dominant and energy levels become more discrete with decreasing system size. In order to estimate the energy level spacing in nanostructures, as a first-order approximation, we consider a free-electron model in a 1D square well potential with perfectly rigid walls as shown in Figure 4.2. By solving the Schrödinger equation, we obtain the average energy level spacing δ at the Fermi level ε_F as

$$\delta \equiv \left(\frac{\pi^2\hbar^2(n+1)^2}{2mL^2}\right) - \left(\frac{\pi^2\hbar^2n^2}{2mL^2}\right) \approx \frac{\pi^2\hbar^2n}{mL^2} = \frac{4\varepsilon_F}{N}, \qquad (4.21)$$

where m is the mass of an electron, $n = 0, 1, 2, \ldots$, and L is the width of the well [2]. If we consider a 3D spherical well potential with a radius of a, a similar argument can be applied and gives the average energy level spacing at ε_F as

$$\delta \sim \frac{1}{\rho(\varepsilon_F)} = \frac{2\varepsilon_F}{3N}, \qquad (4.22)$$

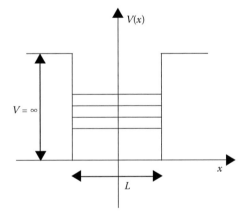

Figure 4.2 1D square well potential with perfectly rigid walls.

where $\rho(\varepsilon_F)$ is the DOS at the Fermi level ε_F [3]. This estimate shows that the energy level spacing becomes larger with decreasing sphere size and is larger than the thermal energy k_BT at low temperatures. When the energy level spacing is larger than the thermal energy, the thermodynamic properties of an electron gas in the sphere are greatly influenced by the discontinuity of the energy level and interesting features appear with significant physical anomalies because the Fermi–Dirac distribution function cannot be used any more.

It should be noted that the above discussion assumes that a free-electron model is a good approximation for the behavior of electrons in normal metals. However, electrons in metals actually correlate with each other and a more detailed discussion must be taken into account in order to examine the detailed electronic structure. On the other hand, the Fermi liquid theory shows that quasiparticle excitation states have a well-defined one-by-one correspondence with the free-electron excitation states [4]. Although the quasiparticle approximation is only used to describe weakly correlated electron systems, here, we adopt the quasiparticle excitations, which are a good first approximation of the electrons considered here, in order to analyze the physical properties arising from low-lying excitation states in nanoparticles.

Next, we discuss the quasiparticle excitation states in a single nanoparticle. Since only one spin-up electron and one spin-down electron can occupy a single quasiparticle excitation state due to the Pauli exclusion principle, the ground state and the low-lying excitation states are depicted as shown in Figure 4.3. Depending on whether the total number of electrons in a

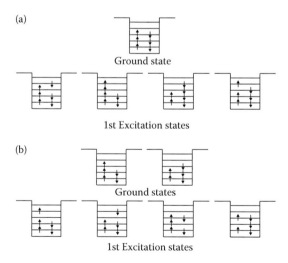

Figure 4.3 Ground states and first quasiparticle excitation states: (a) even and (b) odd.

nanoparticle is even or odd, the excitation states are classified as in the figure and the electronic excitation states of a given single particle can be expressed explicitly in this manner. As an example, we now calculate the electronic specific heat of a nanoparticle. We assume that the total number of electrons is even, and a magnetic field is not applied. By labeling the electronic states of a particle as ... $\varepsilon_{-1}, \varepsilon_0, \varepsilon_1, \varepsilon_2, ...$ from the lower bottom of the energy levels, the ground state is filled with electrons up to the energy level of ε_0, and the lowest-lying excitation states are depicted in Figure 4.3. To calculate the specific heat, write the partition function Z as follows [5]:

$$Z = 1 + 4\exp\left\{\frac{-(\varepsilon_1 - \varepsilon_0)}{k_B T}\right\} + \exp\left\{\frac{-2(\varepsilon_1 - \varepsilon_0)}{k_B T}\right\} + \cdots.$$

(4.23)

Using this partition function, we calculate the free energy and the electronic specific heat,

$$F = -k_B T \log Z,$$

$$C = -T\frac{\partial^2 F}{\partial T^2} = T\frac{\partial^2 (k_B T \log Z)}{\partial T^2}.$$

(4.24)

However, under real experimental conditions, measurements of a single nanoparticle are impossible or extremely difficult; therefore, we usually measure an ensemble of nanoparticles. The ensemble consists of nanoparticles with different sizes and shapes so that the distribution of particle sizes and shapes automatically imposes the need to introduce a distribution function expressing the spectrum of electronic states under consideration, that is, the energy level spacing δ, from a statistical point of view. To do this, we take a theoretical approach based on the random matrix theory proposed by Dyson and Metha [6] to this ensemble of nanoparticles. Although the theory is mathematically very complicated, it tells that the statistical properties of the Hamiltonian operator of a single nanoparticle give very important information about a distribution of the eigenenergy levels of the Hamiltonian operator. In order to obtain the thermodynamic properties of the ensemble of nanoparticles, we need to average Equation 4.24 using the distribution function $P(\varepsilon)$, which expresses the probability of finding the first energy level at an energy level between ε and $\varepsilon + d\varepsilon$. According to the random matrix theory, the distribution function $P(\varepsilon)$ is written as [7]

$$P(\varepsilon) \sim \frac{\varepsilon^n}{\delta^{n+1}},$$

(4.25)

where n is an exponent depending on the magnitude of spin–orbit coupling and applied magnetic field. A general theorem related to n is given below:

(a) Spin–orbit coupling H_{SO} is negligible and no magnetic field H is applied

$$\Rightarrow n=1 \quad (\delta \gg H_{SO}, \ \delta \gg H).$$

(b) Spin–orbit coupling is negligible and large magnetic field is applied

$$\Rightarrow n=0 \quad (\delta \gg H_{SO}, \ \delta \gg H).$$

(c) Spin–orbit coupling is strong and no magnetic field is applied

$$\Rightarrow n=4 \quad (\delta \gg H_{SO}, \ \delta \gg H).$$

(d) Spin–orbit coupling is strong and large magnetic field is applied

$$\Rightarrow n=2 \quad (\delta \gg H_{SO}, \ \delta \gg H).$$

Using this distribution function $P(\varepsilon)$, let us calculate the partition function of the ensemble of nanoparticles:

$$C = T \frac{\partial^2 (k_B T \langle \log Z \rangle_P)}{\partial T^2}, \tag{4.26}$$

where $\langle A \rangle_P$ denotes the average using the distribution function $P(\varepsilon)$, namely

$$\langle \log Z \rangle_P = \int_0^\infty d\varepsilon P(\varepsilon) \ \log(1+4e^{-\varepsilon/k_B T} + e^{-2\varepsilon/k_B T}) = C_1 \left(\frac{k_B T}{\delta} \right)^{n+1}, \tag{4.27}$$

where C_1 is a constant. Therefore, the specific heat C is given by

$$C = C_2 \left(\frac{k_B T}{\delta} \right)^{n+1}. \tag{4.28}$$

Even if the number of electrons is odd, the principal property does not change significantly and only the exponent n varies.

Next, we calculate the magnetic susceptibility. As in the case of specific heat, we first consider a nanoparticle in which the total number of electrons is even and spin–orbit interaction is negligibly small. When a magnetic field H is applied along the z-direction and the energy levels are split by the Zeeman energy of $2\mu_B H$ and $-2\mu_B H$, the partition function is written as

$$Z = 1 + (2 + e^{2\mu_B H} + e^{-2\mu_B H})\, e^{-(\varepsilon_1 - \varepsilon_0)/k_B T} + e^{-2(\varepsilon_1 - \varepsilon_0)/k_B T} + \cdots. \qquad (4.29)$$

This partition function gives the free energy and magnetic susceptibility as

$$\langle F \rangle_p = \langle F(H=0) \rangle_p - C_3 (\mu_B H)^2 \frac{k_B T}{\delta^2}, \qquad (4.30)$$

$$\chi = \frac{2C_3 \mu_B^2 k_B T}{\delta^2}. \qquad (4.31)$$

It should be noted that the magnetic susceptibility obtained here is a factor of (T/δ) smaller than the Pauli susceptibility of metals. If the total number of electrons is odd, only one electron on the Fermi level moves around and this electron gives a Curie-type magnetic susceptibility:

$$\chi = \frac{\mu_B^2}{k_B T}. \qquad (4.32)$$

Other electrons, on the other hand, give a magnetic susceptibility with the same order of magnitude as in the case of an even number of electrons, and the value in Equation 4.32 dominates the total magnetic susceptibility. When the spin–orbit coupling is large enough not to be able to approximate the magnetization as a conservative variable, a direct calculation of the susceptibility using fluctuation dissipation theorem is necessary. Although the details are beyond the content of this textbook, the temperature variation is summarized in Figure 4.4. The most significant point is that the magnetic susceptibility depends on whether the total number of electrons in a particle is even or odd, and this effect is called the Kubo effect. A demonstration of this effect was experimentally reported for Pd clusters with different sizes by Volokitin et al. as shown in Figure 4.5 [8].

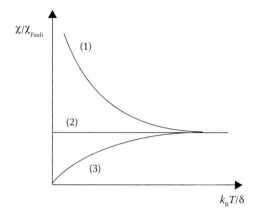

Even or odd	Even	Odd
(a) $\delta \gg H_{SO}$, $\delta \gg H$	(3)	(1)
(b) $\delta \gg H_{SO}$, $\delta \ll H$	(2)	(2)
(c) $\delta \ll H_{SO}$, $\delta \gg H$	(2)	(1)
(d) $\delta \ll H_{SO}$, $\delta \ll H$	(2)	(2)

Figure 4.4 Magnetic susceptibility of nanoparticles as expected from the Kubo effect.

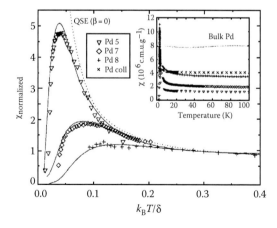

Figure 4.5 Electronic magnetic susceptibility χ_{el} of Pd clusters in the quantum size limit. (Reprinted from Y. Volokitin et al., *Nature* 384, 1996, 621. With permission.)

4.2.2   *Surface magnetism of transition noble metals*

We discussed the size effect of metallic nanoparticles in the previous section, where significant thermodynamic properties appear when the energy level spacing is large compared with the thermal energy. In contrast, the surface region of materials is likely to have a different electronic structure due to the interface effect between the material and vacuum associated with the boundary conditions. It is obvious that a significant surface effect is observed in nanoparticles and thin layers because of their large surface-to-volume ratio. In this section, we will discuss the effect of surfaces on magnetism. In particular, we describe the theoretical and experimental aspects of surface magnetism in 4d transition metals such as Ru, Rh, and Pd, which are paramagnetic in their bulk form. In a theoretical analysis, Blügel studied the surface magnetism of monolayers and bilayers of 4d transition metals and found that 4d transition metals became ferromagnetic due to the surface effect [9]. Such a theoretical prediction was confirmed experimentally in a Ru monolayer on $C(0\,0\,0\,1)$ by Pfandzelter et al. [10].

In order to see what sort of materials can be ferromagnetic, let us briefly review a theoretical description of itinerant magnetism. As a first-order approximation, assume a 3D free-electron gas with the DOS $\rho(\varepsilon_F)$ at ε_F as the itinerant electron system. When we apply a magnetic field along the z-direction, the energy of electrons with spin parallel to the z-direction (spin-up electrons) is raised to a value of $\mu_B H$ and the energy of spin-down electrons decreases by the same magnitude due to Zeeman splitting of the conduction band. Therefore, the number of $\rho(\varepsilon_F)\mu_B H/2$ electrons in the spin-up band at ε_F move to the spin-down band as shown

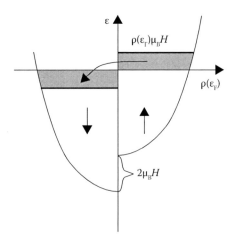

Figure 4.6 Distribution of electrons in a magnetic field of H.

in Figure 4.6, and the system exhibits a magnetization of $2\rho(\varepsilon_F)\mu_B H$. This produces a magnetic susceptibility:

$$\chi = 2\mu_B^2\rho(\varepsilon_F), \tag{4.33}$$

which is called the Pauli paramagnetic susceptibility [3].

What happens if exchange coupling between electrons is included in the calculation of magnetic susceptibility? To proceed to this calculation, we denote the numbers of spin-up and spin-down electrons in a metallic system as $N_\uparrow$ and $N_\downarrow$, respectively. Since a pair of itinerant electrons with antiparallel spin coupling gains an exchange energy I, to a first approximation, the total exchange energy E_{ex} is given as

$$E_{ex} = IN_\downarrow N_\uparrow = \frac{1}{4}IN^2 - \frac{1}{4}IM^2, \tag{4.34}$$

where N is the total number of itinerant electrons, and $M = N_\downarrow - N_\uparrow$ is the magnetization of the electron gas with a unit of μ_B. On the other hand, a change in the total kinetic energy due to the exchange splitting is given by

$$E_{kin} - E_{kin}^0 = \int_0^{\varepsilon_F - \Delta} \varepsilon\rho(\varepsilon)\,d\varepsilon + \int_0^{\varepsilon_F + \Delta} \varepsilon\rho(\varepsilon)\,d\varepsilon - 2\int_0^{\varepsilon_F} \varepsilon\rho(\varepsilon)\,d\varepsilon, \tag{4.35}$$

where $\rho(\varepsilon_F)$ is the DOS and Δ is the exchange splitting. Equation 4.35 can be expanded as a series of Δ as

$$E_{kin} - E_{kin}^0 = \rho(\varepsilon_F^0)\Delta^2 \left\{ 1 - \frac{3}{4}\left(\frac{\rho'(\varepsilon_F^0)^2}{\rho(\varepsilon_F^0)^2} - \frac{\rho''(\varepsilon_F^0)}{3\rho(\varepsilon_F^0)} \right)\Delta^2 + \cdots \right\}. \tag{4.36}$$

If we assume conditions that the total number of electrons is constant and independent of the value of Δ, then Δ is expressed in a series of M. Using this relation between Δ and M, Equation 4.36 is rewritten as

$$E_{kin} - E_{kin}^0 = \frac{1}{4\rho(\varepsilon_F^0)}M^2 + \frac{1}{64\rho(\varepsilon_F^0)^3}\left(\frac{\rho'(\varepsilon_F^0)^2}{\rho(\varepsilon_F^0)^2} - \frac{\rho''(\varepsilon_F^0)}{3\rho(\varepsilon_F^0)} \right)M^4 + \cdots. \tag{4.37}$$

As a consequence, the total gain of exchange energy and kinetic energy ΔE is given as

$$\Delta E = \frac{1}{4\rho(\varepsilon_F^0)}\{1 - I\rho(\varepsilon_F^0)\}M^2 + \frac{1}{64\rho(\varepsilon_F^0)^3}\left(\frac{\rho'(\varepsilon_F^0)^2}{\rho(\varepsilon_F^0)^2} - \frac{\rho''(\varepsilon_F^0)}{3\rho(\varepsilon_F^0)} \right)M^4 + \cdots. \tag{4.38}$$

From this equation, we obtain the condition for spontaneous magnetization to occur: the total energy gain must be minimized at a finite $M \neq 0$ when

$$1 - I\rho(\varepsilon_F^0) < 0. \tag{4.39}$$

This condition is called the Stoner criteria for ferromagnetism of itinerant electron systems [11]. Equation 4.38 immediately gives the corresponding magnetic susceptibility as

$$\chi = \frac{2\rho(\varepsilon_F^0)}{1 - I\rho(\varepsilon_F^0)}. \tag{4.40}$$

It should be noted that the magnetic susceptibility is enhanced by a factor of $1/(1 - I\rho(\varepsilon_F))$ compared with the Pauli paramagnetic susceptibility calculated above. This factor is called the Stoner enhancement factor and it becomes large when the DOS at ε_F is large or even shows a divergence at $I\rho(\varepsilon_F) = 1$. Thus, from this discussion, we intuitively conclude that if the electronic structure varies and the DOS becomes large or has a peak at the Fermi energy as the system size is decreased in nanoparticles or thin films, then the system, which is paramagnetic in bulk form, becomes ferromagnetic because of the variations in the electronic structure due to the surface effect. This is the basic idea of how to induce magnetism in metallic nanostructures.

In order to observe ferromagnetic behaviors in nanostructures, 4d transition metals are suitable because the DOS has a peak at or near the Fermi energy in bulk, thus the Stoner criteria is fulfilled due to the modification of the electronic structure by size effects and/or surface effects. Here are some examples of ferromagnetism in nanostructures. Zhu et al. [12] calculated the electronic structures of Pd and Rh thin layers deposited on Ag and Au substrates using an *ab-initio* approach and found ferromagnetic behavior in Rh, but no ferromagnetism of Pd was predicted in their results. Eriksson et al. [13] also carried out a theoretical analysis of Ru, Rh, and Pd layers on Ag substrates and obtained similar results. The reason for the absence of ferromagnetism in Pd was explained in terms of a possible hybridization of 4d orbitals with the substrate due to the relatively expanded 4d band of Pd. However, Blügel's analysis of Pd predicted a significant size dependence of electronic structures of Pd [9]. A summary of Blügel's calculation is given in Table 4.1, where a bilayer of Pd shows ferromagnetism, whereas a monolayer remains paramagnetic. Also, a bilayer was reported to exhibit a peak in the DOS at the Fermi energy and the peak to move below the Fermi level in the monolayer. In this case, the Stoner criteria for ferromagnetism are fulfilled in the bilayer,

Table 4.1 Magnetic Moments on 4*d* Transition
Metal on Ag in the Unit of Bohr Magneton

	Pd	Rh	Ru
ML on Ag	0	1.04	1.73
BL on Ag	0.20	0.32	0.05

thereby inducing ferromagnetism in Pd. This mechanism is in principle different from that of other 4*d* transition metals, that is, the combined effects of both surface effect and bulk-like properties are necessary for ferromagnetism of Pd.

For structures of nanoparticles and clusters, Reddy et al. calculated the electronic structures of Ru_{13}, Rh_{13}, and Pd_{13} clusters and found them all to show ferromagnetism [14]. Similar results were also reported in a calculation by Piveteau et al. [15] and Lee [16]. In terms of the size dependence of magnetism in clusters, Guirado-López et al. [17] calculated the size dependence of the magnetism of Ru and Ph clusters and Lee [18] did the same for Pd clusters. They obtained the surprising results that the magnetic moment per atom reduces with increasing cluster size. Moreover, the symmetry of the cluster was taken into account in the calculation by Vitos et al. [19], and they found an icosahedral cluster of Pd to be more suited than the octahedoral symmetry to satisfy the criteria for ferromagnetism.

Experimentally, the above predictions have been tested in recent years because of a significant advance in technology for fabricating nanostructures. Although some results indicated there to be an absence of ferromagnetism in 4*d* transition metals, Pfandzelter et al. [10] reported ferromagnetism of Ru monolayers on highly oriented pyrolytic graphites using spin polarized photoemission experiments. In Figure 4.7, clear ferromagnetic signals are observed below 250 K. In this case, the use of C(0 0 0 1) substrates was very important for this observation because mutual interdiffusion between the monolayer and substrate was greatly suppressed and the electronic structure of C(0 0 0 1) was significantly different from that of Ru, and thus the hybridization of orbitals was greatly reduced. For clusters, Cox et al. [20] observed superparamagnetic properties of Rh clusters in Stern–Gerlach experiments and showed a clear size dependence of the magnetic moments. Recently, Shinohara et al. [21] also found a surprising result of ferromagnetism in free-standing Pd nanoparticles, where only the (100) facet shows ferromagnetism, which agrees with the theoretical predictions above. Also a similar ferromagnetic behavior was reported in chemically prepared Au nanoparticles [22].

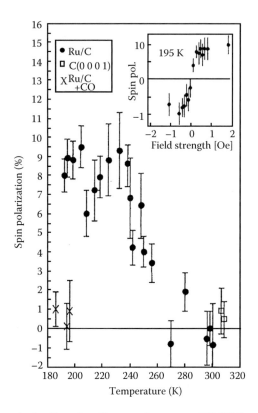

Figure 4.7 Spin polarization of an Ru monolayer on a C(0 0 0 1) substrate detected by spin polarized secondary electron spectroscopy. (Reprinted from R. Pfandzelter, G. Steierl, and C. Rau, *Phys. Rev. Lett.* 74, 1995, 3467. With permission.)

4.2.3   *Single-domain structures and superparamagnetism*

Ferromagnetic materials in which magnetic moments of atoms align parallel to each other usually have domain structures in order to decrease the total magnetostatic energy, and two adjacent domains are separated by a domain wall as shown in Figure 4.8. In this section, we calculate the

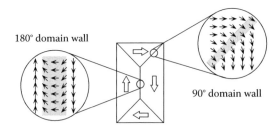

Figure 4.8 Domain structure of a ferromagnetic material.

total magnetic energy of a ferromagnet and describe the types of magnetic domain structures that occur when the size of the ferromagnet is decreased to a little less than a few micrometers. Let us assume that the total energy of the magnet is given by the sum of the magnetic anisotropy energy E_A, exchange energy E_{exch}, and magnetostatic energy E_m. First, we derive a formula for the magnetic anisotropy energy E_A. When the magnetization of a material M_s is orientated along the direction cosine $(\alpha_1, \alpha_2, \alpha_3)$, then, generally, the magnetic anisotropy energy can be expanded in terms of a series of α_1, α_2, and α_3 as

$$E_A = K_0 + K_1(\alpha_1^2\alpha_2^2 + \alpha_2^2\alpha_3^2 + \alpha_3^2\alpha_1^2) + K_2\alpha_1^2\alpha_2^2\alpha_3^2 + \cdots, \quad (4.41)$$

where K_0, K_1, and K_2 are called the anisotropy constants. If we take a polar coordinate (θ, ϕ), Equation 4.41 is rewritten as

$$E_A = K_1 \sin^2\theta + \left\{\frac{1}{4}(K_1 + K_2)\sin^2\phi - K_1\right\}\sin^4\theta - \frac{K_2}{4}\sin^2 2\phi \sin^6\theta + \cdots, \quad (4.42)$$

using $\alpha_1 = \sin\theta \cos\phi$, $\alpha_2 = \sin\theta \sin\phi$, $\alpha_3 = \cos\theta$. Also, we rewrite this as

$$E_A = \frac{K_1}{64}\{(3 - \cos 2\theta + \cos 4\theta)(1 - \cos 4\phi) + 8(1 - \cos 4\theta)\}$$
$$+ \frac{K_2}{256}\{1 - \cos 4\phi(2 - \cos 2\theta - 2\cos 4\theta + \cos\theta)\} + \cdots \quad (4.43)$$

Now, we model a very simple domain structure with a 180° domain wall as shown in Figure 4.9. If the domain wall consists of $(N + 1)$ lattice planes, and the relative angle ϕ between spins S in the nearest neighbor

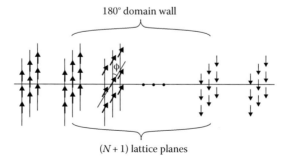

Figure 4.9 A 180° domain wall consisting of $(N + 1)$ lattice planes.

planes is equal to π/N in the domain wall, then the total exchange energy per unit area of the domain wall is written as

$$E_{\text{exch}} = -2J \sum_{i<j} \mathbf{S}_i \cdot \mathbf{S}_j \approx \frac{JS^2\pi^2}{Na^2} = \frac{JS^2\pi^2}{\delta a}, \tag{4.44}$$

where the sum is taken for all the pairs of $\mathbf{S}_i$ and $\mathbf{S}_j$ in the unit area of the domain wall, J is the exchange constant, and δ is the width of the domain wall. If we approximate the magnetic anisotropy energy E_A as the first-order K_1 term in Equation 4.42, then E_A is expressed in a uniaxial form

$$E_A = K_1 \sin^2 \theta. \tag{4.45}$$

Using this uniaxial anisotropy, we can calculate the magnetic anisotropy energy of the domain wall in the unit area as

$$E_A = K_1 a \sum_{n=0}^{N} \sin^2 n\phi = K_1 a \sum_{n=0}^{N} \sin^2\left(\frac{n\pi}{N}\right). \tag{4.46}$$

Let us now transform this sum into the corresponding integral form

$$E_A = K_1 a \int_0^N \sin^2\left(\frac{n\pi}{N}\right) dn = \frac{1}{2} K_1 \delta. \tag{4.47}$$

From Equations 4.44 and 4.47, the total energy of the domain wall is obtained as

$$E_W = E_A + E_{\text{exch}} = \frac{JS^2\pi^2}{\delta a} + \frac{1}{2} K_1 \delta. \tag{4.48}$$

In order to minimize the total energy E_W, the width of the domain wall and the domain wall energy are written as

$$\delta = \left(\frac{2JS^2\pi^2}{aK_1}\right)^{1/2} = \sqrt{2}\pi \sqrt{\frac{A}{K_1}}, \tag{4.49}$$

$$E_W = \sqrt{2}\pi \sqrt{AK_1},$$

where $A = JS^2/a$ is the exchange stiffness constant. An important point is that the domain wall width δ decreases with increasing magnetic anisotropy.

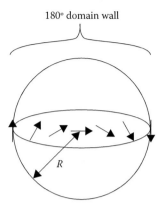

180° domain wall

R

Figure 4.10 Critical size of a single-domain nanoparticle.

Let us now see what happens when the size of the magnet becomes very small compared with the domain wall width. We can intuitively infer that the domain wall energy is larger than the anisotropy energy and the ferromagnet cannot be divided into a multidomain structure any more and becomes a single domain. This situation can be easily confirmed. Assume a spherical nanoparticle in which the magnetic moments rotate by 180° from the left-hand side edge to the right as shown in Figure 4.10. This means that the whole particle consists of a domain wall and this is the smallest particle size with a domain wall. The domain wall energy of this particle U_W is written using Equations 4.2 through 4.27 as

$$U_W \approx \left(\frac{JS^2\pi^2}{Ra} + \frac{1}{2} K_1 R \right) \pi R^2. \tag{4.50}$$

On the other hand, if the particle does not have any domain walls and all the magnetic moments are parallel to each other, then the total magneto-static energy is written as

$$U_m = \frac{1}{6} M^2 \left(\frac{4}{3} \pi R^3 \right). \tag{4.51}$$

Therefore, $U_W = U_m$ gives the condition of the critical size R determining whether the particle has a domain wall and a multidomain structure or a single-domain one. The resulting critical size is given as

$$R = \sqrt{\frac{2JS^2\pi^2}{a(4M^2/9 - K_1)}}. \tag{4.52}$$

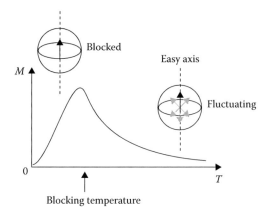

Figure 4.11 Blocking phenomena of a magnetic nanoparticle with uniaxial magnetic anisotropy.

When a ferromagnetic nanoparticle has a single-domain structure, let us see what happens to its magnetic properties. The first example is a blocking phenomenon. Since all magnetic moments in a single nanoparticle rotate coherently, the total magnetic moment of the particle can be treated as if it were a huge atomic moment. Given the energy of the magnetic moments as a function of the orientation in a uniaxial magnetic anisotropy as shown in Figure 4.11, the coherent magnetic moment stays at either of the energy minima *A* or *B* at low temperatures. However, when thermal energy becomes larger than the energy barrier between the two minima, then the direction of the magnetic moment can pass over the barrier and the magnetic moment rotates as a huge paramagnetic moment. This situation is called superparamagnetism. In contrast, when the condition that the energy barrier is higher than the thermal energy is satisfied, the magnetic moment starts to be blocked in certain directions with decreasing temperature. This temperature is called blocking temperature. Since the magnetic moment is fixed at a certain easy direction below the blocking temperature, the magnetization decreases. This situation is illustrated in Figure 4.11. Under these conditions, if the magnetization reversal occurs by thermal excitation over the barrier, then the relaxation of the magnetization $M(t)$ is given as

$$M(t) = M_S \exp\left(\frac{-t}{\tau}\right), \tag{4.53}$$

where τ is the relaxation time of the magnetization and written as

$$\tau = \tau_0 \exp\left(\frac{VK_1}{k_B T}\right), \tag{4.54}$$

where V is the volume of the particle and the parameter τ is the inverse of the attempt frequency, which usually has a value in the range 10^{-9}–10^{-13} s. From this equation, the blocking temperature can be calculated when the magnetic anisotropy constant and the volume of the particle are known. In the above discussion, we attribute the magnetic relaxation process to be thermally activated; however, tunneling through the energy barrier also contributes to the relaxation process, in particular, at very low temperatures. We will discuss the tunneling process in the next section.

4.3 Ferromagnetic domain-wall-related phenomena

We have observed some examples of interesting phenomena appearing in a small magnet due to size and surface effects. Also, it is seen that ferromagnetic materials have a domain structure that depends on the system size. In this section, we focus on other magnetic and magnetotransport characteristics of ferromagnets, in particular, domain-wall-related phenomena. In recent years, such phenomena are of interest because of recent progress in nanofabrication technology such as lithography and thin film growth.

4.3.1 Macroscopic quantum tunneling in magnetic nanostructures

Before discussing macroscopic quantum tunneling in magnetic nanostructures, let us review macroscopic quantum tunneling of a particle with a mass of m in the double well potential shown in Figure 4.12a. Classical mechanics tells that at $T = 0$ the particle stays at around the local minimum in the well. In quantum mechanical limit, however, the particle can oscillate between the two local minima due to tunneling processes at a frequency of $\omega = \Delta E_0/\hbar$, where ΔE_0 is the tunneling splitting of the

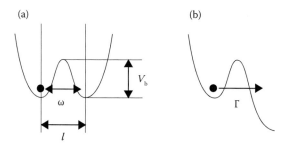

Figure 4.12 (a) Macroscopic quantum coherence and (b) macroscopic quantum tunneling.

ground state. If the potential is that shown in Figure 4.12b, on the other hand, the particle put in the potential minimum at time $t = 0$ will escape from the potential at a mean escape rate Γ due to quantum fluctuation. In the semiclassical limit $\hbar \rightarrow 0$ or $m \rightarrow \infty$, Γ is given as

$$\Gamma \sim \omega_0 \exp\left(-\frac{1}{\hbar}\sqrt{mV_b}\, l\right), \tag{4.55}$$

where ω_0 is the frequency of the classical particle motion around the local minimum, l is the distance between the minima, and V_b is the potential barrier height.

Analogous to the case of the particle, let us consider a tunneling process of the spin degree of freedom. We assume that the Hamiltonian operator of a single spin S is given by

$$H = -A(S^x)^2 + B(S^z)^2 - g\mu_B HS^y, \tag{4.56}$$

where A and B are constants. In the classical limit, the Hamiltonian operator is reduced to

$$H(\theta, \phi) = -AS^2 \cos^2 \phi \sin^2 \theta + BS^2 \cos^2 \theta - g\mu_B H \sin \theta \sin \phi, \tag{4.57}$$

where θ and ϕ are the polar and azimuthal angles, respectively (Figure 4.13a). When no magnetic fields are applied, both the positive and negative directions along the x-axis are equal energy minima for the spin, that is, $\theta = \pi/2$, $\phi = 0$ or π, and the landscape of the energy potential is similar to the double well potential in Figure 4.12a. Therefore, a transition between $\phi = 0$ and π at $\theta = \pi/2$ can be considered as a quantum tunneling process of the spin (see Figure 4.13b). To discuss the macroscopic quantum tunneling of the spin in more detail, we need Feynman's path integral approach

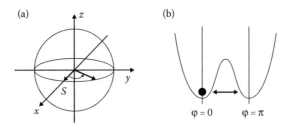

Figure 4.13 (a) Classical ground states of magnetization in $H = 0$ and $H \neq 0$ and (b) landscape of the energy potential.

[23]. Although this approach is beyond the scope of this textbook, the calculation gives the tunneling splitting for $H = 0$ as

$$\Delta E_0 = \frac{16}{\sqrt{\pi}} S^{3/2} \left[\frac{A}{B} \left(1 + \frac{A}{B} \right) \right]^{3/4} \left[\sqrt{1 + \frac{A}{B}} + \sqrt{\frac{A}{B}} \right]^{-2S}. \tag{4.58}$$

The result means that there is a finite quantum spin tunneling probability even at $T = 0$ for a large macroscopic spin.

Next, we present some experimental demonstrations of macroscopic quantum tunneling for magnetization. Now, a single crystal array of Mn_{12}-ac is a typical material showing macroscopic quantum tunneling behavior. Mn_{12}-ac crystals consist of dodecanuclear $[Mn_{12}(H_3COO)_{16}(H_2O)_4 O_{12}] \cdot 2CH_3COOH \cdot 4H_2O$ molecules, water of crystallization, and disordered acetic acid molecules. The presence of Mn ions produces a ferromagnetic ground state and each Mn_{12} molecule in the tetragonal crystal has a total spin $S = 10$. The magnetization of Mn_{12}-ac shows clear superparamagnetic behavior with a single relaxation time of $\tau = \tau_0 \exp(KV/k_BT)$. If we measure the magnetic relaxation of Mn_{12} molecules, then the relaxation time shows the temperature dependence of $\log(\tau/\tau_0) \propto 1/T$ at high temperatures, while in the low-temperature regime, the relaxation time has a plateau, independent of temperature (see Figure 4.14) [24]. In the high-temperature regime, the magnetic relaxation is due to thermal excitation over the energy barrier between two local minima of the magnetic anisotropy. In the low-temperature regime, such thermal relaxation ceases to occur and quantum tunneling across the barrier becomes predominant, leading to the plateau of the relaxation time. This is a clear signature of macroscopic quantum tunneling

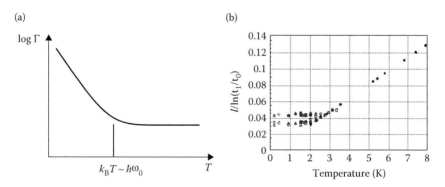

Figure 4.14 (a) Schematic diagram of crossover from thermal activation to quantum tunneling and (b) experimental demonstration using magnetic nanoparticles. (Reprinted from B. Barbara et al., *J. Magn. Magn. Mater.* 140–144, 1995, 1825. With permission.)

(a)   (b)

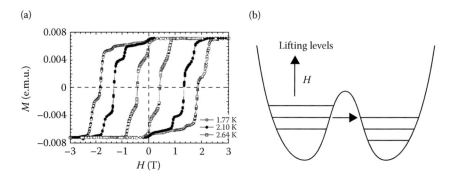

Figure 4.15 Macroscopic quantum tunneling of magnetization. (Reprinted from L. Thomas et al., *Nature* 383, 1996, 145. With permission.)

of magnetization. Macroscopic quantum tunneling between discrete energy levels of an Mn_{12} cluster is also observed in the magnetization curve as shown in Figure 4.15a [25]. Since the energy level is lifted with increasing magnetic field, macroscopic quantum tunneling of magnetization becomes significant when the neighboring energy levels match each other; the situation is shown in Figure 4.15b. Recently, such nanoclusters are considered as possible quantum bits—qubits—for use in quantum computers—because quantum coherence reaches 100 μs as demonstrated in Rabi oscillations of V_{15} clusters anion [26]. Therefore, magnetic clusters may find wide range of applications in advanced electronic devices.

Another interesting object that exhibits macroscopic quantum tunneling is a magnetic domain wall as shown in Figure 4.16. A domain wall

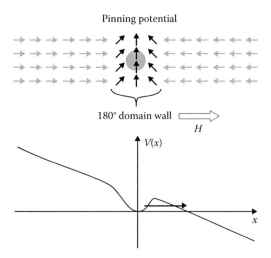

Figure 4.16 Domain wall pinned at a potential minimum.

is pinned in a fixed potential minimum, and in the classical limit the domain wall moves to another nearest potential minimum over the potential barrier due to thermal excitation. On the other hand, such thermally excited movement is suppressed at very low temperatures, that is, in the quantum limit, the movement of a domain wall occurs by a tunneling process through the energy barrier.

4.3.2 Electron scattering at domain walls: Quantum coherence

In the previous section, we saw that ferromagnetic materials have a domain structure, where neighboring domains with different directions of magnetic moments are separated by domain walls. The width of a domain wall is determined by the exchange constant and magnetic anisotropy constant of the material. Now, we discuss the electron transport properties in ferromagnetic materials and, in particular, describe what happens when conduction electrons are transmitted across a domain wall, namely, how the domain walls scatter conduction electrons. Since the resistivity of materials is attributed to electron scattering, what we discuss here is the resistivity arising from domain walls, namely, whether domain walls increase or decrease the resistivity. In order to analyze electron transport in ferromagnetic materials, we use a two-current model where spin-up and spin-down current channels carry electrons with each of the spins existing in the ferromagnet. Since a domain wall is a region with spatially rotating spins, conduction electron spins may not be able to follow the local spins in the domain wall in traversing the domain wall, that is, electron spins move across domain walls nonadiabatically. In a ferromagnet, on the other hand, one of the two conduction spin channels—called the majority spin channel—has larger conductivity and predominantly contributes to electron conduction. Therefore, if electrons in the majority spin channel are scattered at a domain wall into the minority low-conductive spin channel owing to nonadiabaticity, then the domain wall gives an additional contribution to the resistivity. Levy and Zhang [27] calculated this effect by a semiclassical approach and obtained the domain wall contribution for current parallel to the domain walls ρ_{CIW} as

$$\rho_{CIW} = \rho_0 \left\{ 1 + \frac{\xi^2}{5} \frac{(\rho_0^{\uparrow} - \rho_0^{\downarrow})^2}{\rho_0^{\uparrow} \rho_0^{\downarrow}} \right\},$$

$$\xi = \frac{\hbar v_F}{J\delta},$$

(4.59)

where ρ_0 is the resistivity without domain walls, $\rho_0^{\uparrow}$ and $\rho_0^{\downarrow}$ are the resistivities for spin-up and spin-down channel electrons, the parameter ξ is

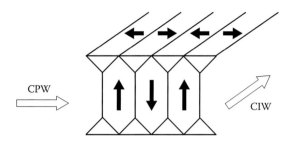

Figure 4.17 Current direction for CIW and CPW configurations.

a measure of the nonadiabaticity, J is the exchange energy, and v_F is the Fermi velocity (Figure 4.17). For current perpendicular to the domain walls, the resistivity ρ_{CPW} is obtained as

$$\rho_{CPW} = \rho_0 \left\{ 1 + \frac{\xi^2}{5} \frac{(\rho_0^\uparrow - \rho_0^\downarrow)^2}{\rho_0^\uparrow \rho_0^\downarrow} \left(3 + \frac{10\sqrt{\rho_0^\uparrow \rho_0^\downarrow}}{\rho_0^\uparrow + \rho_0^\downarrow} \right) \right\}. \tag{4.60}$$

From these two equations, we obtain the ratio of ρ_{CPW} and ρ_{CIW} as

$$\frac{\rho_{CPW}}{\rho_{CIW}} = 3 + \frac{10\sqrt{\rho_0^\uparrow \rho_0^\downarrow}}{\rho_0^\uparrow + \rho_0^\downarrow}. \tag{4.61}$$

Equation 4.61 enables a simple test of the magnitude of the resistivity contribution that could be observed in experiments, given typical parameters.

In contrast to Levy and Zhang's prediction, Tatara and Fukuyama treated the domain wall contribution to the resistivity in the weak localization limit and saw completely opposite behavior, that is, domain walls were found to reduce the resistivity [28]. In the linear response theory, quantum interference of electrons is significant in a disordered ferromagnet and induces weak localization of electrons, which increases the resistivity. However, in a domain wall, electron coherence is almost totally destroyed so that a weak localization is likely to be suppressed, suggesting that a domain wall contributes to a reduction in the resistivity. Provided that l is the elastic mean free path, L is the wire length, and L_∞ is the wire width, in the limit $L/l \gg 1$ and $\kappa \equiv \tau/\tau_W \ll 1$, where τ is the lifetime due to normal impurity scattering and τ_W is the lifetime due to a domain wall, the quantum correction of conductivity σ_Q due to a domain wall is obtained as

$$\sigma_Q = \frac{e^2 n\tau}{m} \frac{6}{k_F^2 L_\perp^2} \left\{ \frac{L}{l} - \frac{\tan^{-1}(\sqrt{3\kappa}L/l)}{\sqrt{3\kappa}} \right\}, \tag{4.62}$$

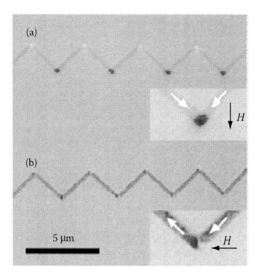

Figure 4.18 **(See color insert following page 148.)** Different domain wall configurations of ferromagnetic zigzag wires after applying a magnetic field (a) perpendicular and (b) parallel to the horizontal axis.

where n is the electron density. This value can be compared with experimental observations using typical physical parameters.

In order to check whether domain walls increase or decrease the resistivity, a number of experimental studies have been carried out using ferromagnetic metal and ferromagnetic semiconductor wires with well-engineered magnetic domain structures. Kent et al. [29] demonstrated the fabrication of bcc Fe, hcp Co, and $L1_0$ FePt wires and found a novel negative contribution to the resistivity at low temperatures in Fe wires. Taniyama et al. [30] controlled the domain structures of zigzag-shaped Co wires and found a similar result. On the other hand, Gregg et al. [31] found a positive contribution to the resistivity in a Co epitaxial layer, a result that is opposite to the previous results. Oxide materials such as $SrRuO_3$ were also used to gain insight into this domain wall effect. An example of a method for detection of domain wall resistivity is shown in Figure 4.18.

4.3.3 Spin current and spin transfer torque–current-induced domain wall motion

Spin current is one of the most important concepts for understanding the physical phenomena governing spin electronics such as spin transfer torque [32,33]. Here, we first introduce the meaning of spin current. In the

s–d model of magnetism, the Hamiltonian operator of a spin system can be written as

$$H = -\sum_{\sigma}\sum_{i,j} t_{ij} c_{i\sigma}^{+} c_{j\sigma} - 2J \sum_{i} s_i \cdot S_i,\qquad(4.63)$$

where t_{ij} is the transfer integral, $c_{i\sigma}^{+}$ and $c_{i\sigma}$ are the creation and annihilation operators of an electron, and s_i and S_i are the spins of an itinerant electron and a localized electron of the ith atom, respectively [11]. From Equation 4.63, we can derive the equation of motion for s_i and S_i using the Heisenberg equation to be

$$\frac{\mathrm{d}}{\mathrm{d}t}\langle s_i \rangle = \frac{i}{\hbar}\langle[H,s_i]\rangle = -\frac{1}{2}\nabla \cdot j_i^{S} + \frac{2}{\hbar}J\langle s_i \times S_i \rangle$$

$$\frac{\mathrm{d}}{\mathrm{d}t}\langle S_i \rangle = \frac{i}{\hbar}\langle[H,S_i]\rangle = -\frac{2}{\hbar}J\langle s_i \times S_i \rangle,\qquad(4.64)$$

where j_i^{S} is the spin current operator, which can be expressed as

$$j_i^{S} = -\frac{i}{\hbar}\sum_{j}\sum_{\sigma\sigma'} R_{ij}\{(-t_{ji})c_{j\sigma}^{+}(2s)_{\sigma\sigma'}c_{i\sigma'} - (-t_{ij})c_{i\sigma}^{+}(2s)_{\sigma\sigma'}c_{j\sigma'}\}.\qquad(4.65)$$

Since the total magnetic moment M_i of the ith atom is written as $M_i = 2(s_i + S_i)$ in units of μ_B, the equation of motion for M_i is

$$\frac{\mathrm{d}}{\mathrm{d}t}M_i = -\nabla \cdot j_i^{S}.\qquad(4.66)$$

This corresponds to the law of conservation of electron spin angular momentum and also can be intuitively understood using the illustration in Figure 4.19 [31,32]. When a spin current J_{in} is injected into a magnet

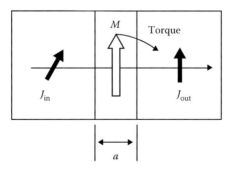

Figure 4.19 Spin current and spin transfer torque.

having a magnetization M and J_{out} is ejected from the magnet, Equation 4.66 is reduced to

$$\frac{\mathrm{d}}{\mathrm{d}t}M = -\frac{\delta J^S}{a}, \qquad (4.67)$$

where $\delta J^S = J_{out} - J_{in}$ is the change in the spin current. From this equation, we expect that M is exerted a torque. The torque is called the spin transfer torque. Let us see an example of this effect for the case when electron current flows within a ferromagnetic wire with a domain wall as depicted in Figure 4.20. Since the domain wall is much longer than the Fermi wavelength of an electron, the direction of spin current j_i^S is parallel to the magnetization direction; in other words, electron spin rotates following the magnetization direction in the domain wall in an adiabatic limit, and to a good approximation, the spin current can be written as

$$j_i^S = P e_i \frac{j^e}{e}, \qquad (4.68)$$

where e_i is the unit vector of the magnetic moment, j^e is the electron current, and P is the spin polarization of the material defined as

$$P = \frac{(\sigma_\uparrow - \sigma_\downarrow)}{(\sigma_\uparrow + \sigma_\downarrow)}, \qquad (4.69)$$

where $\sigma_\uparrow$ and $\sigma_\downarrow$ are the conductivities of each spin channel. Assuming that in this electron transmission process the absolute value of the

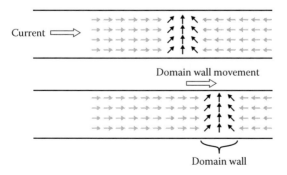

Figure 4.20 Current-induced domain wall movement.

magnetization M does not change and that just the orientation of M changes, we rewrite Equation 4.66 as

$$\frac{d}{dt} M_i = -\nabla \cdot j_i^S = -\frac{Pj^e}{e}\frac{de_i}{dz},$$

$$\frac{d\theta}{dt} = -\frac{Pj^e}{eM}\frac{d\theta}{dz}, \tag{4.70}$$

where θ is the orientation of magnetization with respect to the wire axis and given with an arbitrary function $f(z)$ as

$$\theta = f\left(z - \frac{Pj^e}{eM}t\right). \tag{4.71}$$

The important point from this solution is that domain walls move toward the direction along which electron current flows in magnetic wires. The movement of domain walls has been observed experimentally in submicron ferromagnetic wires as shown in Figure 4.21 [34]. In this experiment, current flowing in a submicron NiFe wire pushes a domain wall at a velocity of 3 m/s, which is consistent with the theoretical prediction given above.

If we apply this spin transfer torque effect to magnetic multilayers as shown in Figure 4.22, we can deduce that the switching of magnetization is possible using spin current. In order to observe this effect, suppose

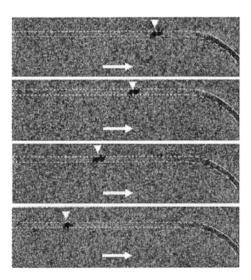

Figure 4.21 **(See color insert following page 148.)** Experimental demonstration of current-induced domain wall motion. (Reprinted from A. Yamaguchi et al., *Phys. Rev. Lett.* 92, 2004, 077205. With permission.)

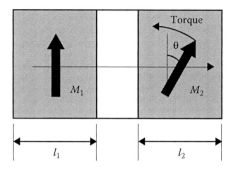

Figure 4.22 Current-induced magnetization switching.

that a nonmagnetic layer is sandwiched between two ferromagnetic layers with magnetizations of M_1 and M_2 with layer widths of l_1 and l_2 and that the direction of these magnetizations makes an angle θ. By applying Equation 4.66, we obtain equations of motion for the magnetizations of the two ferromagnetic layers to be

$$\frac{d}{dt}M_1 = \frac{j^e/l_1}{P_2^{-1}+\cos\theta}\{\hat{M}_1\times(\hat{M}_1\times\hat{M}_2)\}$$

$$\frac{d}{dt}M_2 = \frac{j^e/l_2}{1+P_2\cos\theta}\{\hat{M}_2\times(\hat{M}_1\times\hat{M}_2)\}.$$

(4.72)

From Equation 4.72, both magnetizations exerted a torque as shown in Figure 4.23 and the switching of magnetization is realized using spin

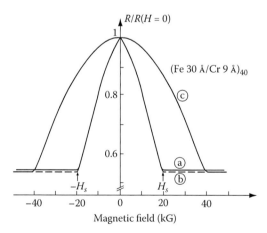

Figure 4.23 GMR in multilayers of Fe/Cr for the current in-plane and the field in-plane (a, b), or perpendicular to the plane (c). (Reprinted from M.N. Biabich et al., *Phys. Rev. Lett.* 61, 1988, 2472. With permission.)

current. A theory for this mechanism was first proposed by Berger [32] and Slonczewski [33] and experimentally the spin current-induced magnetization switching was observed in magnetic trilayers such as Co/Cu/Co structures [35].

4.4 Spin transport in magnetic nanostructures: Magnetic interface effect

The most prominent issue in magnetic nanostructures is giant MR (GMR) in magnetic multilayers discovered by Grünberg et al. [36] and Biabich et al. in Fert's research group [37]. Since the discovery, world-wide studies on spin-dependent transport have led to GMR being used as the principal mechanism of magnetic sensors in read heads of magnetic information media over the last 10 years. In this section, we first introduce the mechanism of spin-dependent transport that underpins the emerging field of spintronics in magnetic multilayers and magnetic tunnel junctions. An understanding of spin-dependent electron transport across heterogeneous interfaces has also led to an expansion of research in this area, and other novel concepts such as spin accumulation and spin Hall effects have been discovered. These topics are also reviewed in the following section.

4.4.1 GMR and TMR effect: Spin-dependent scattering in multilayers and tunneling junctions

GMR was first discovered in ferromagnetic Fe/nonmagnetic Cr multilayers as shown in Figure 4.23. The essence of GMR is that the resistance of a magnetic multilayer depends on the relative orientation of the magnetization of adjacent ferromagnetic layers, that is, when the relative orientation of the ferromagnetic layers changes from antiparallel to parallel, then the resistance decreases and saturates at a field where the magnetizations of these two layers are completely aligned in parallel. Let us consider the basic phenomenological theory of GMR. Here, we define spin-up ($\uparrow$) and spin-down ($\downarrow$) with respect to the fixed z-axis, that is, the direction of magnetic field, while majority and minority spins are defined as spin+ and spin−. Consider a multilayer consisting of equally thick $2n$ layers as shown in Figure 4.24. If we employ a two-current model as discussed before, the total resistance ρ_P for parallel magnetization alignment can be written as

$$\frac{1}{\rho_P} = \frac{1}{\rho_\uparrow} + \frac{1}{\rho_\downarrow},$$

$$\rho_\uparrow = 2n\rho_+,$$

$$\rho_\downarrow = 2n\rho_-,$$

(4.73)

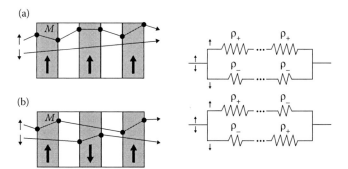

Figure 4.24 Spin-dependent scattering in magnetic multilayers for (a) parallel and (b) antiparallel magnetization configurations. Scattering centers are depicted with circles.

where ρ_+ and ρ_- are the resistance of majority and minority spin channels, respectively. For antiparallel magnetic alignment, on the other hand, the resistance is given as

$$\frac{1}{\rho_{AP}} = \frac{1}{\rho_{\uparrow}} + \frac{1}{\rho_{\downarrow}},$$

$$\rho_{\uparrow} = \rho_{\downarrow} = n(\rho_+ + \rho_-). \qquad (4.74)$$

Therefore, we can calculate the MR as being

$$MR = \frac{\rho_{AP} - \rho_P}{\rho_{AP}} = \left(\frac{\rho_+ - \rho_-}{\rho_+ + \rho_-}\right)^2. \qquad (4.75)$$

Equation 4.75 indicates that a finite MR is expected when $\rho_+ \neq \rho_-$ is known. This situation can be easily understood when we write an equivalent circuit as shown in Figure 4.24. However, experimentally reported results are not consistent with the above theoretical prediction, so we need another scenario in order to explain the experiments. According to a theory by Inoue, they pointed out that the predominant origin of GMR in Fe/Cr multilayers is due to the spin-dependent DOS of Fe and Cr, as shown in Figure 4.25 [38]. Since the DOS of Fe for spin-up states is located below the Fermi level, only spin-up electrons of the Cr layer are scattered into the Fe layers at the interface while realizing the scattering potential V. The relaxation times $\tau_{\uparrow}$ and $\tau_{\downarrow}$ of spin-up and spin-down electrons are given as

$$\frac{1}{\tau_{\uparrow\downarrow}} = \sum_{k'} |\langle k'|V_{\uparrow\downarrow}|k\rangle|^2 \delta(\varepsilon_k - \varepsilon_{k'}), \qquad (4.76)$$

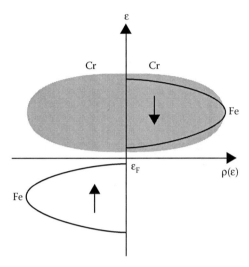

Figure 4.25 Schematic diagram of the DOS of Fr and Cr.

where k and k' are the wavevectors of incident and scattered electrons, respectively, and MR can be close to 100% and a large MR is expected.

Soon after the discovery of GMR, the phenomenon of tunnel MR (TMR) was reported by Miyazaki and Tezuka [39] and Moodera et al. [40]. A typical TMR structure consists of two ferromagnetic layers with a nonmagnetic insulating interlayer such as in $Fe/Al_2O_3/Fe$ structures. The experimental results reported by Miyazaki et al. are shown in Figure 4.26, where the resistance increases when the relative orientation of magnetization is antiparallel, similar to GMR. Let us consider the phenomenological theory called Jullire's model, which is useful to characterize the TMR ratio. According to Landauer's formula, it is widely known that the conductance Γ of the tunnel junction shown in Figure 4.27 is written as

$$\Gamma = \frac{e^2}{\hbar} \sum_{k_\parallel, k_\parallel', \sigma} T_\sigma(\varepsilon_F, k_\parallel, k_\parallel'), \qquad (4.77)$$

in the limit of bias voltage $V \to 0$ [41], where T_σ is the tunneling probability of spin-σ electrons across the insulating layer, and $k_\parallel$ and $k_\parallel'$ are the wavevectors parallel to the layer plane. Assuming that the wavevectors $k_\parallel$ and $k_\parallel'$ are independent, then in this approximation, Equation 4.77 is reduced to

$$\Gamma \propto \frac{e^2}{\hbar} \sum_{\sigma} D_{1\sigma}(\varepsilon_F) D_{2\sigma}(\varepsilon_F) T, \qquad (4.78)$$

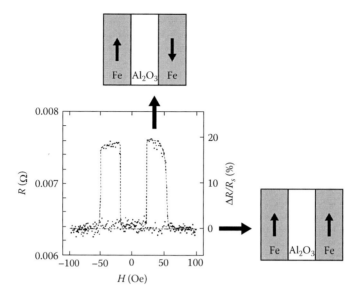

Figure 4.26 TMR curve of an Fe/Al$_2$O$_3$/Fe trilayer. (Reprinted from T. Miyazaki and T. Tezuka *J. Magn. Magn. Mater.* 139, 1995, L231. With permission.)

where $D_{1\sigma}$ and $D_{2\sigma}$ are the DOS of spin-σ electrons in the magnetic layers 1 and 2, respectively. This equation gives the conductance for parallel and antiparallel magnetic alignments as

$$\Gamma_P \propto D_{1\uparrow}(\varepsilon_F)D_{2\uparrow}(\varepsilon_F) + D_{1\downarrow}(\varepsilon_F)D_{2\downarrow}(\varepsilon_F),$$
$$\Gamma_{AP} \propto 2D_{1\uparrow}(\varepsilon_F)D_{2\downarrow}(\varepsilon_F). \tag{4.79}$$

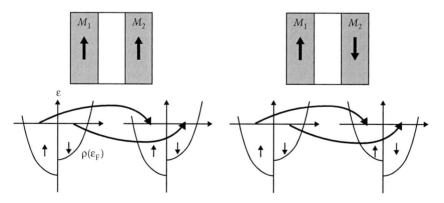

Figure 4.27 Tunneling process across a TMR junction.

As a consequence, MR is obtained using the spin polarization of the two electrodes P_1 and P_2 as follows:

$$MR \equiv \frac{\Gamma_{AP}^{-1} - \Gamma_P^{-1}}{\Gamma_{AP}^{-1}} = \frac{2P_1P_2}{1+P_1P_2}. \qquad (4.80)$$

This equation is called Jullire's formula and is widely used to estimate the MR ratio of magnetic tunnel junctions. In the above discussion, we took the limit of $V \to 0$. However, for a finite voltage, inelastic scattering effects become more significant, namely, magnon scattering, spin-flip scattering at magnetic impurities, and so on. These effects are important, in particular, to develop real TMR devices. To include these effects, it is worthwhile showing Landauer's formula in a finite voltage,

$$\Gamma = \frac{e}{\hbar} \int d\varepsilon \{ f(\varepsilon) - f(\varepsilon + eV) \} \sum_{k_\parallel, k_\parallel', \sigma} T_\sigma(\varepsilon_F, k_\parallel, k_\parallel'), \qquad (4.81)$$

where $f(\varepsilon)$ is Fermi–Dirac's distribution function. Although the above discussion gives the basics of TMR in magnetic tunnel junctions, it should be noted that a recent *ab initio* calculation by Butler et al. predicted that a huge MR up to 1000% can be expected when MgO(001) is used as a tunneling barrier in Fe/MgO(001)/Fe tunnel junctions [42]. One of the most important points in this calculation is that the Δ_1 band of Fe hybridizes well with the orbital of MgO along the axis from Γ point to H point in reciprocal space, and therefore, a quasi 1D path from Fe into MgO carries the conduction electrons. The second point is that the Δ_1 band for majority spin electrons has a finite DOS at the Fermi level, while that for minority spin electrons has no DOS as shown in Figure 4.28. This indicates that

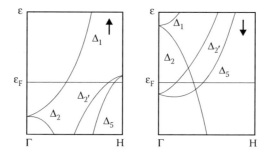

Figure 4.28 Energy dispersion of bcc-Fe for spin-up and spin-down states.

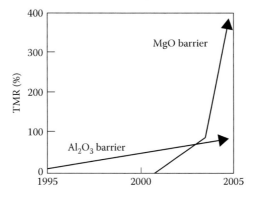

Figure 4.29 Trends of the development of TMR ratio.

spin transport through the Δ_1 band gives a huge MR according to Jullire's model. Moreover, since there are no minority spin bands that can hybridize with any orbital of MgO at $k_\parallel = (0, 0)$, electrons with a finite $k_\parallel$ inevitably contribute to the tunneling conduction and the conduction electron channel decays when transmitting across the MgO tunneling barrier. This means that the MgO works as a spin filter. This is the fundamental physics predicted by Butler et al. and such a combination of factors in Fe/MgO/Fe tunnel junctions produces huge changes in MR. Experimentally, Yuasa et al. [43] confirmed the theoretical predictions for Fe/MgO/Fe tunnel junctions and Butler's calculation was supported by experiments. This prediction also provided a tremendous improvement in the magnitude of TMR as shown in Figure 4.29.

4.4.2 *Spin accumulation and current-perpendicular-to-plane (CPP) GMR: Spin diffusion length*

The concept of spin accumulation at magnetic metal/nonmagnetic metal interfaces produces fascinating spin transport properties in CPP GMR. In order to understand this concept, we start with the equation of motion for magnetization given by Equation 4.66. According to a theory by Zhang et al., they considered only the z components of magnetization and spin current for simplicity in this equation, and introduced the relaxation term of spin as

$$\frac{d}{dt}M = -\nabla \cdot j_z^S - \frac{\delta M}{\tau}, \tag{4.82}$$

where δM is the deviation of magnetization from the average value. The z component of the spin current density j_z^S and total electron current density j^e can be written as

$$j_z^S = j_\uparrow^e - j_\downarrow^e,$$
$$j^e = j_\uparrow^e + j_\downarrow^e, \tag{4.83}$$

using spin-up and spin-down electron current densities $j_\uparrow^e$ and $j_\downarrow^e$. Taking into account the diffusion process of electron spin, $j_\uparrow^e$ and $j_\downarrow^e$ are given as

$$j_{\uparrow(\downarrow)}^e = \sigma_{\uparrow(\downarrow)}E(x) - D_{\uparrow(\downarrow)}\nabla(\delta n_{\uparrow(\downarrow)}), \tag{4.84}$$

where $\sigma_{\uparrow(\downarrow)}$, $D_{\uparrow(\downarrow)}$, and $\delta n_{\uparrow(\downarrow)}$ are the conductivities, the diffusion constant, and the deviation of charge from the average value for spin-up or spin-down electrons, respectively. From Equations 4.82 through 4.84, we obtain j^e and j_z^S as

$$j^e = (\sigma_\uparrow + \sigma_\downarrow)E(x) - \frac{D_\uparrow - D_\downarrow}{2}\frac{\partial}{\partial x}\delta M - \frac{D_\uparrow + D_\downarrow}{2}\frac{\partial}{\partial x}\delta n,$$
$$j_z^S = (\sigma_\uparrow - \sigma_\downarrow)E(x) - \frac{D_\uparrow + D_\downarrow}{2}\frac{\partial}{\partial x}\delta M - \frac{D_\uparrow - D_\downarrow}{2}\frac{\partial}{\partial x}\delta n. \tag{4.85}$$

Assuming that $\delta n = 0$, the effective electric field $E(x)$ is given by

$$E(x) = \frac{j^e}{(\sigma_\uparrow + \sigma_\downarrow)} + \frac{D_\uparrow - D_\downarrow}{2(\sigma_\uparrow + \sigma_\downarrow)}\frac{\partial}{\partial x}\delta M,$$
$$j_z^S = \frac{\sigma_\uparrow - \sigma_\downarrow}{\sigma_\uparrow + \sigma_\downarrow}j^e + \frac{\sigma_\uparrow D_\downarrow + \sigma_\downarrow D_\uparrow}{\sigma_\uparrow + \sigma_\downarrow}\frac{\partial}{\partial x}\delta M = Pj^e - \xi\nabla(\delta M), \tag{4.86}$$

where P is the spin polarization and ξ is the averaged diffusion constant defined in the equation. By substituting the expression of j_z^S in Equation 4.85, we obtain the diffusion equation for δM,

$$\frac{d}{dt}M - \xi\nabla^2\delta M + \frac{\delta M}{\tau} = -j^e\nabla P. \tag{4.87}$$

Now, let us see the x dependence of δM for two typical magnetic configurations, that is, when a nonmagnetic layer is sandwiched between two ferromagnetic layers and the ferromagnetic layers are aligned (a) parallel or (b) antiparallel as shown in Figure 4.30. When we pass electron current j^e from the left-hand side ferromagnetic layer to the right-hand side layer

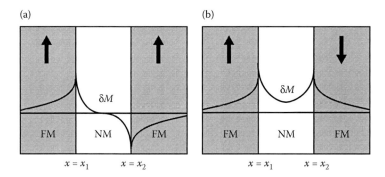

Figure 4.30 Spin accumulation at FM/NM interfaces for (a) parallel and (b) antiparallel configurations.

across the nonmagnetic layer, the stationary solution near the interface of $x = x_1$ is given by

$$\delta M(x) = \frac{j^e P \lambda_{F(N)}}{2\xi} \{e^{(x-x_1)/\lambda_F}\theta(x_1 - x) + e^{-(x-x_1)/\lambda_N}\theta(x - x_1)\},$$

$$\lambda_F = \sqrt{\xi \tau_F},$$ (4.88)

$$\lambda_N = \sqrt{D_N \tau_N},$$

where λ_F and λ_N, which can be expressed with spin relaxation times τ_F and τ_N, and diffusion constants ξ and D_N, correspond to the spin diffusion constants in the ferromagnetic and nonmagnetic layers. This situation is illustrated in Figure 4.30. Also, δM for condition (b) can be calculated as shown in Figure 4.30b. From these results, electron spins accumulate at the FM/NM interfaces when electron current flows across the ferromagnet/nonmagnet interface. The δM is called spin accumulation. Of interest is that the sign of the spin accumulation switches as the current direction is reversed. Let us consider the contribution of spin accumulation to the resistance. Since a diffusion current is included in j^e of Equation 4.84, an additional resistance ΔR occurs due to spin accumulation as

$$\Delta R = \frac{1}{j^e A} \int_{-\infty}^{\infty} dx \left\{ E(x) - \frac{j^e}{\sigma_\uparrow + \sigma_\downarrow} \right\} = \frac{P_0}{2A} \frac{D_\uparrow - D_\downarrow}{\sigma_\uparrow + \sigma_\downarrow} \sqrt{\frac{\tau_F}{\xi}},$$ (4.89)

using Equations 4.85 and 4.88, where A is the area of the interface. This additional MR was first recognized by Valet and Fert [44].

Spin accumulation can also be discussed on the basis of an electrochemical potential at FM/NM metal interfaces. Let us formulate spin accumulation at the interface with this approach. We start with the continuity condition of the total current density and spin current density for spin-up and spin-down channels as follows:

$$\frac{d}{dx}(j_\uparrow^e + j_\downarrow^e) = 0,$$

$$\frac{d}{dx}j_{\uparrow(\downarrow)}^e = -\frac{e}{2}\left(\frac{\delta n_{\uparrow(\downarrow)}}{\tau_{\uparrow(\downarrow)}} - \frac{\delta n_{\downarrow(\uparrow)}}{\tau_{\downarrow(\uparrow)}}\right),$$

(4.90)

where $\tau_\uparrow$ and $\tau_\downarrow$ are the spin flip time for spin-up and spin-down electrons. Using the general relationships

$$\delta n_{\uparrow(\downarrow)} = \delta\mu_{\uparrow(\downarrow)}\rho_{\uparrow(\downarrow)}(\varepsilon_F), \qquad (4.91)$$

$$\frac{\rho_{\uparrow(\downarrow)}(\varepsilon_F)}{\tau_{\uparrow(\downarrow)}} = \frac{\rho_{\downarrow(\uparrow)}(\varepsilon_F)}{\tau_{\uparrow(\downarrow)}}: \quad \text{equilibrium relationship,} \qquad (4.92)$$

$$D_{\uparrow(\downarrow)} = \frac{\sigma_{\uparrow(\downarrow)}}{e^2\rho_{\uparrow(\downarrow)}(\varepsilon_F)}: \quad \text{Einstein's equation,} \qquad (4.93)$$

we obtain equations for the electrochemical potential:

$$\nabla^2(\sigma_\uparrow\mu_\uparrow + \sigma_\downarrow\mu_\downarrow) = 0,$$

$$\nabla^2(\mu_\uparrow - \mu_\downarrow) = \frac{1}{\lambda_S^2}(\mu_\uparrow - \mu_\downarrow),$$

(4.94)

where $\lambda_S = (\xi\tau_{sf})^{1/2}$ is the spin diffusion length, ξ is the effective diffusion constant, and τ_{sf} is the effective spin flip time given by $1/\tau_{sf} = (\tau_\uparrow^{-1} + \tau_\downarrow^{-1})/2$. Since Equation 4.94 is the diffusion equation, it can be solved as

$$\mu_\uparrow(x) = C_1 + C_2 x + \frac{C_3}{\sigma_\uparrow}e^{-x/\lambda_S} + \frac{C_4}{\sigma_\uparrow}e^{x/\lambda_S},$$

$$\mu_\downarrow(x) = C_1 + C_2 x - \frac{C_3}{\sigma_\downarrow}e^{-x/\lambda_S} - \frac{C_4}{\sigma_\downarrow}e^{x/\lambda_S}.$$

(4.95)

Thus, the spin accumulation δM, which is proportional to $(\mu_\uparrow - \mu_\downarrow)$, can be obtained from this approach based on the electrochemical potential and is consistent with that given in Equation 4.88. The spin accumulation is observed experimentally using a nonlocal spin valve device as shown in Figure 4.31. The effect was first demonstrated by Jedema et al. [45] and discussed by a number of research groups.

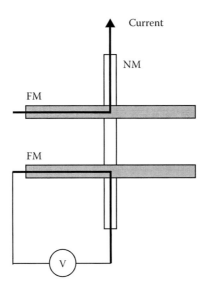

Figure 4.31 Nonlocal detection of spin accumulation at an FM/NM interface.

4.4.3 Spin Hall effect: Side jump and skew scattering due to spin–orbit coupling

When an electric current j_x is flowing through a nonmagnetic conducting wire in a magnetic field H perpendicular to the current direction x, electrons feel the Lorentz force and move toward the direction y perpendicular to both the current and magnetic field. These electrons finally accumulate at the edge of the wire and electric field E_y results across the wire. This is called the Hall effect and the Hall resistance E_y/j_x is given by R_0H using the Hall coefficient $R_0 = -1/(nec)$. The Hall resistance of ferromagnetic materials, on the other hand, has an additional term $4\pi R_s M$ and written as

$$R_H = R_0H + 4\pi R_s M, \tag{4.96}$$

where M is the magnetization and R_s is called the extraordinary Hall coefficient. Obviously, the first term is the ordinary Hall resistance arising from the Lorentz force and the second term is the extraordinary term proportional to the magnetization [46]. The extraordinary term is mainly due to spin–orbit coupling in the materials. There are two explicit origins of the extraordinary Hall effect, one of which is intrinsic and due to the atoms in the crystal, and the other is extrinsic, arising from spin–orbit scattering around impurities. Since most spin Hall effects in magnetic nanostructures are originated extrinsically, here, let us consider the fundamental mechanism of spin–orbit scattering due to its extrinsic origin, and start with the spin–orbit coupling Hamiltonian operator written as

$$H_{SO} = \zeta \{\nabla V_{imp}(r) \times p\} \cdot \sigma = \zeta \{\sigma \times \nabla V_{imp}(r)\} \cdot p, \qquad (4.97)$$

where ζ is the coupling constant, V_{imp} is the impurity potential, and p and σ are the momentum and spin of an electron, respectively. From this Hamiltonian operator, the velocity of electrons is given by

$$v = \frac{\partial H_{SO}}{\partial p} = \frac{p}{m} + \zeta \{\sigma \times \nabla V_{imp}(r)\}. \qquad (4.98)$$

The second term of Equation 4.98 means that an additional shift of the velocity occurs depending on the spin of the electron under consideration. This is called a side jump. On the other hand, the first term of Equation 4.98 also gives rise to another extraordinary Hall effect contribution when the impurity potential contains spin–orbit coupling. This effect is due to electron scattering depending on the orientation of electron spin and is called skew scattering. These two mechanisms are extrinsic in origin and give rise to the extraordinary Hall effect as illustrated in Figure 4.32. The main point here is that spin-up and spin-down electrons deflect oppositely when flowing in a material. Since the numbers of spin-up and spin-down electrons are different in ferromagnetic materials, excess majority electrons accumulate at the edge of samples. Therefore, the excess electrons give rise to an additional Hall voltage. This was an intuitive description of the extraordinary Hall effect.

Next, we discuss methods for determining which particular mechanisms predominate in experiments. Generally, the Hall resistivity is measured as

$$\rho_S = (\sigma^{-1})_{xy} = \frac{-\sigma_{xy}}{\sigma_{xx}^2 + \sigma_{xy}^2} \cong \frac{-\sigma_{xy}}{\sigma_{xx}^2}. \qquad (4.99)$$

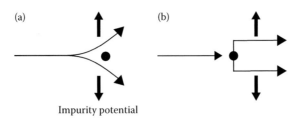

(a)   (b)

Impurity potential

Figure 4.32 Extrinsic origins of extraordinary Hall effect: (a) skew scattering and (b) side jump mechanisms.

On the other hand, a microscopic theory according to the Kubo formula shows that the orthogonal term of the conductivity σ_{xx} is given by

$$\sigma_{xx} = e^2 \sum_{\sigma} \frac{n_{\sigma}}{m} \tau_{\sigma}, \qquad (4.100)$$

for both mechanisms, and is called the Drude formula. For the nonorthogonal term of the conductivity, the side jump and skew scattering mechanisms give different contributions σ_{xy}^{SJ} and σ_{xy}^{SS} as

$$\sigma_{xy}^{SJ} = -2e^2 \varsigma m,$$

$$\sigma_{xy}^{SS} = -2\pi \frac{e^2 \varsigma}{\hbar} \Omega_0 V_{imp} \sum_{\sigma} \frac{1}{2} n_{\sigma} \tau_{\sigma}, \qquad (4.101)$$

where Ω_0 is the volume of a unit cell. The important fact from Equations 4.101 is that σ_{xy}^{SJ} is independent of the relaxation time τ_{σ}, whereas σ_{xy}^{SS} is proportional to τ_{σ}. Therefore, the Hall resistivity due to a side jump is proportional to $(1/\tau_{\sigma}^2) \sim \rho_{xx}^2$ but that arising from skew scattering is proportional to $(1/\tau_{\sigma}) \sim \rho_{xx}$. Checking this relationship in experiments, we can determine the predominant origin of the extraordinary Hall effect in a specific system.

Next, consider the Hall effect in nonmagnetic materials. Because the number of spin-up and spin-down electrons is the same in this case, no anomalous Hall voltage occurs across the sample. Even so, the same number of electrons with different spins accumulates at the opposite edges of a sample due to spin–orbit coupling, causing a spin current (Figure 4.33).

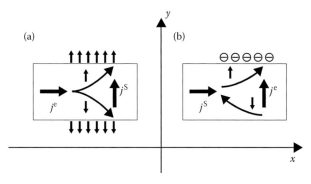

Figure 4.33 Schematic diagram of the (a) spin Hall effect and (b) inverse spin Hall effect.

This effect is called the spin Hall effect [47]. The total spin current j_x^S can be written as

$$j_x^S = -\frac{\sigma_N}{e}\frac{\partial}{\partial x}(\delta\mu_N) + \sigma_{xy}^N E_y, \qquad (4.102)$$

where σ^N is the conductivity of the nonmagnetic material, $\delta\mu_N = (\mu_N^\uparrow - \mu_N^\downarrow)/2$, and $\sigma_{xy}^N = (\sigma_{xy}^\uparrow - \sigma_{xy}^\downarrow)$. This expression shows that both the side jump and skew scattering mechanisms give rise to spin current along the y-direction and electron spins accumulate at both edges. We can also infer that the inverse case, where spin current induces charge current perpendicular to the spin current direction, is also possible. This situation is called the inverse spin Hall effect as depicted in Figure 4.33b. The point to note about the inverse Hall effect is that spin current is the difference between charge current with spin-up $j^\uparrow$ and that with spin-down $j^\downarrow$, that is, $j^S = j^\uparrow - j^\downarrow$. When spin current flows along the y-direction, then charge currents $j^\uparrow$ and $j^\downarrow$ with different spins deflect oppositely. Therefore, the

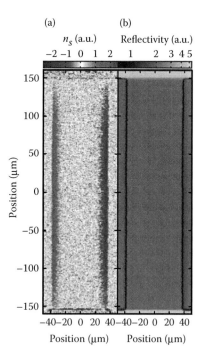

Figure 4.34 (**See color insert following page 148.**) Experimental confirmation of the spin Hall effect. (Reprinted from Y.K. Kato et al., *Science* 306, 2004, 1910. With permission.)

total charge current $j^e = j^\uparrow + j^\downarrow$ flows along the x-direction. Similar to the spin current, the corresponding charge current is written as

$$j_x^e = \sigma_N E_x + \frac{\sigma_{xy}^N}{\sigma_N} j_y^S. \tag{4.103}$$

The spin Hall effect [48] and the inverse spin Hall effect [49] have been experimentally confirmed as shown in Figure 4.34.

PROBLEMS

4.1 Derive Equations 4.7 and 4.11 and check that both approaches give the same exchange energy term.

4.2 Derive Equation 4.37 using the conditions given in the text.

4.3 Which material is suitable for preparing a large single-domain particle? Give the reason.

4.4 The temperature-dependent ac zero-field-cooled magnetization of an ensemble of single-domain ferromagnetic particles shows a peak at T_B as a consequence of the blocking phenomenon of each magnetic moment. Show the measurement frequency dependence of T_B.

4.5 If you want to observe the resistivity due to a ferromagnetic domain wall in a nonadiabatic limit, what is the appropriate choice of a ferromagnetic material?

4.6 Derive Equation 4.64.

4.7 Estimate the typical velocity of domain wall motion in a ferromagnetic wire with a spin polarization of 40% and a magnetization of 1500 emu/cm^3 with an electric current of 10^6 A/cm^2 flowing through the wire.

4.8 To achieve a high TMR ratio, what sort of materials are appropriate for the two ferromagnetic layers?

4.9 As a general means to detect spin accumulation in a nonmagnetic metal, the approach of nonlocal detection is usually employed as shown in Figure 4.31. Show the nonlocal voltage due to spin accumulation as a function of magnetic field.

4.10 If you want to detect the spin Hall effect clearly, what could be good choices for materials?

References

1. R. Kubo, *J. Phys. Soc. Jpn.* 17, 1962, 976.
2. L.I. Schiff, *Quantum Mechanics*, 3rd Ed., McGraw-Hill, Inc., Auckland, 1968.
3. N.W. Ashcroft and N.D. Mermin, *Solid State Physics*, Thomson Learning, Inc., Australia, 1976.
4. P.W. Anderson, *Basic Notions of Condensed Matter Physics*, Benjamin-Cummings Publishing Company, Inc., London, 1984.

5. R.P. Feynman, *Statistical Mechanics, A Set of Lectures*, Addison-Wesley Publishing Company, Inc., Massachusetts, 1972.
6. Dyson and M.L. Metha, *Random Matrices and Statistical Theory of Energy Levels*, Academic Press, New York and London, 1967.
7. A. Kawabata, *Solid State Phys.* 7, 1972, 29.
8. Y. Volokitin, J. Sinzig, L.J. de Jongh, G. Schmid, M.N. Vargaftik, and I.I. Moiseev, *Nature* 384, 1996, 621.
9. S. Blügel, *Phys. Rev. B* 51, 1995, 2025.
10. R. Pfandzelter, G. Steierl, and C. Rau, *Phys. Rev. Lett.* 74, 1995, 3467.
11. K. Yoshida, *Magnetism*, Springer-Verlag, New York, 1995.
12. M.J. Zhu, D.M. Byander, and L. Kleinman, *Phys. Rev. B.* 43, 1991, 4007.
13. O. Eriksson, R.C. Albers, and A.M. Boring, *Phys. Rev. Lett.* 66, 1991, 1350.
14. B.V. Reddy, S.N. Khanna, and B.I. Dunlap, *Phys. Rev. B* 51, 1995, 13852.
15. B. Piveteau, M.C. Desjpnquères, A.M. Olés, and D. Spanjaard, *Surf. Sci.* 352–354, 1996, 951.
16. K. Lee, *Z. Phys. D* 40, 1997, 164.
17. R. Guirado-López, D. Spanjaard, and M.C. Desjpnquères, *Phys. Rev. B* 57, 1998, 6305.
18. K. Lee, *Phys. Rev. B* 58, 1998, 2391.
19. L. Vitos, B. Johansson, and J. Kollár, *Phys. Rev. B* 62, 2000, R11957.
20. A.J. Cox, J.G. Louderback, and L.A. Bloomfield, *Phys. Rev. Lett.* 71, 1993, 923.
21. T. Shinohara, T. Sato, and T. Taniyama, *Phys. Rev. Lett.* 91, 2003, 197201.
22. Y. Yamamoto, T. Miura, M. Suzuki, N. Kawamura, H. Miyagawa, T. Nakamura, K. Kobayashi, T. Teranishi, and H. Hori, *Phys. Rev. Lett.* 93, 2004, 116801.
23. L. Gunther and B. Barbara, *Quantum Tunneling of Magnetization—QTM'94*, Kluwer Academic Publishers, Dordrecht, 1995.
24. B. Barbara, W. Wernsdorfer, L.C. Sampaio, J.G. Park, C. Paulsen, M.A. Novak, R. Ferr et al., *J. Magn. Magn. Mater.* 140–144, 1995, 1825.
25. L. Thomas, F. Lionti, R. Ballou, D. Gatteschi, R. Sessoli, and B. Barbara, *Nature* 383, 1996, 145.
26. S. Bertaina, S. Gambarelli, T. Mitra, B. Tsukerblat, A. Muller, and B. Barbara, *Nature* 453, 2008, 203.
27. P.M. Levy and S. Zhang, *Phys. Rev. Lett.* 79, 1997, 5110.
28. G. Tatara and H. Fukuyama, *Phys. Rev. Lett.* 78, 1997, 3773.
29. A.D. Kent, J. Yu, U. Rudiger, and S.S.P. Parkin, *J. Phys.: Condens. Matter.* 13, 2001, R461.
30. T. Taniyama, I. Nakatani, T. Namikawa, and Y. Yamazaki, *Phys. Rev. Lett.* 82, 1999, 2780.
31. J.F. Gregg, W. Allen, K. Ounadjela, M. Viret, M. Hehn, S.M. Thompson, and M.D.J. Coey, *Phys. Rev. Lett.* 77, 1996, 1580.
32. L. Berger, *J. Appl. Phys.* 49, 1978, 2156.
33. J.C. Slonczewski, *J. Magn. Magn. Mater.* 159, 1996, L1.
34. A. Yamaguchi, T. Ono, S. Nasu, K. Miyake, K. Mibu, and T. Shinjo, *Phys. Rev. Lett.* 92, 2004, 077205.
35. E.B. Myers, D.C. Ralph, J.A. Katine, R.N. Louie, and R.A. Buhrman, *Science* 285, 1999, 867.
36. P. Grünberg R, Schreiber, Y. Pang, M.B. Brodsky, and H. Sowers, *Phys. Rev. Lett.* 57, 1986, 2442.
37. M.N. Biabich J.M. Broto, A. Fert, F. Nguyen Van Dau, F. Petroff, P. Eitenne, G. Creuzet, A. Friederich, and J. Chazelas, *Phys. Rev. Lett.* 61, 1988, 2472.

38. J. Inoue, *J. Mag. Soc. Jpn.* 25, 2001, 1384, 1448, 1484.
39. T. Miyazaki and T. Tezuka, *J. Magn. Magn. Mater.* 139, 1995, L231.
40. J.S. Moodera, L.R. Kinder, T.M. Womg, and R. Meservey, *Phys. Rev. Lett.* 74, 1995, 3273.
41. S. Datta, *Electronic Transport in Mesoscopic Systems*, Cambridge University Press, Cambridge, 1995.
42. W.H. Butler, X.-G. Zhang, T.C. Schulthess, and J.M. MacLaren, *Phys. Rev. B* 63, 2001, 054416.
43. S. Yuasa, T. Nagahama, A. Fukushima, Y. Suzuki, and K. Ando, *Nat. Mater.* 3, 2004, 868.
44. T. Valet and A. Fert, *Phys. Rev. B* 48, 1993, 7099.
45. F.J. Jedema, A.T. Filip, and B.J. van Wees, *Nature* 410, 2001, 345.
46. A. Crepieux and P. Bruno, *Phys. Rev. B* 64, 2001, 014416.
47. J.E. Hirsh, *Phys. Rev. Lett.* 83, 1999, 1834.
48. Y.K. Kato, R.C. Myers, A.C. Gossard, and D.D. Awschalom, *Science* 306, 2004, 1910.
49. E. Saitoh, M. Ueda, H. Miyajma, and G. Tatara, *Appl. Phys. Lett.* 88, 2006, 182509.

Index

A

Absorption of light, 131
Acceptance molecules
    electron-exchange energy transfer, 188
    photo-induced electron transfer, 190
Acenes absorption spectra, 142
Aggregate
    excited state, 150
    molecular linear chain, 152
Alkanethiol monolayer, 103
Aluminum
    magic cluster stability, 119
    magic numbers, 118
Aluminum oxide development, 245
Amplitude, 7
Angle-resolved ultraviolet photoelectron
        spectroscopy (ARUPS), 71
    experimental setup and principle, 72
Angular momentum quantum number, 83
Anisotropic energy, 116
Antibunching phenomenon, 158
Antiferromagnetism, 208
Aromatic systems, 140
Arsenic
    quantum dots, 195
    quantum wells, 196
Artificial atoms, 79
ARUPS. *See* Angle-resolved ultraviolet
        photoelectron spectroscopy
        (ARUPS)
Avogadro's number, 184

B

Band dispersion, 93
Band structure graphical representation, 47
Bessel differential equation, 84

Bessel function, 84
Bloch function, 35, 36, 150
    solids electronic band structured, 33
Blügel's analysis of lead, 222
Body-centered lattice, 41, 43
Bohr radius, 139
    excited and ground states, 190
    Rydberg exciton energy, 136
Bohr, Niels, 7
Bohr–Sommerfeld condition, 68, 93, 94–95
Bra-ket notation, 18–19, 21
Bragg's diffraction, 31, 32
Brillouin function, 212
Brillouin zone, 35, 42, 43, 47, 48, 49, 65
    graphite layer, 49
Building blocks, 58
Bulk-forbidden band gap, 98
Bulk-localized electronics states, 97

C

C-centered lattice, 41
Cadmium selenium quantum dot
    antibunching phenomenon, 158
    fluorescence time trace, 158, 159
    time delays histogram, 158
Channels, 74
Charge-transfer exciton, 134
Charging effect, 101
    nanosphere, 102
Charging energy, 101
Chemical potential. *See* Fermi level
Chromium, 242
    GMR, 239
    torque, 239
Coherent length of exciton, 151
Commutable, 8
Commute, 8

Complex refractive index, 162
Concept of probability, 7
Conductance
    nanowire, 75, 78
    quantization, 78
Confinement potentials
    electrons, 90
    origin, 90–93
Conjugated oligomers, 143
Conjugated organic systems, 143
Conjugated polymers, 141
Copenhagen interpretation, 7
Coulomb binding energy, 135
Coulomb blockade, 102
    experimental setup, 102
    gold nanoparticles, 103
Coulomb force, 172
Coulomb interaction, 51, 133, 210
Coulomb staircase, 103
CPP. *See* Current-perpendicular-to-plane
    (CPP)
Crystalline materials
    bulk, 98
    periodic boundary conditions, 65
    size, 65
    surface energy, 116
Crystalline silver substrate, 82
Curie's law, 212
Curie-type magnetic susceptibility, 218
Curie–Weiss law, 207
    ferromagnetic spin system, 213
Current
    CIW and CPW configurations, 234
    magnitude calculation, 74
Current-induced domain wall motion,
    237, 238
Current-induced magnetization
    switching, 239
Current-perpendicular-to-plane (CPP),
    245–248
Cylindrical well two dimensional
    construction, 85
    quantum confinement, 83–84

D

Damping constant, 199
Dangling bonds, 98
Dansyl-(L-prolyl)-alpha naphthyl
    chemical structure, 185
    energy transfer as function of
        distance, 186
    fluorescence excitation spectra, 185

De Broglie, 12
Delocalization length of exciton, 151
Density functional method, 54–55
Density of states (DOS). *See* Electron
    density of states
Dexter energy transfer, 187–189
Dexter exchange mechanism, 190
Dexter, D. L., 187
Diatomic molecule, 97
Dielectric constants, 133
Dielectrical surface, 200
Dimer
    absorption transition probabilities, 149
    energy levels change, 146
    ground state, 144
    Hamiltonian operator, 144
    molecular interaction, 146
    Schrödinger equation, 144
Dipole–dipole interaction, 151
    J-aggregate spectrum, 154
    resonant energy transfer, 180
Dirac's bra–ket notation, 18–19, 21
Dirac's delta function, 131
Dirac, Paul, 18
Direct band gap, 136
Domain wall
    lattice planes, 225
    pinned at potential minimum, 232
Donor molecules
    electron-exchange energy transfer, 188
    photo-induced electron transfer, 190
DOS. *See* Electron density of states
Double-slit experiment, 1
    using bullets, 2
    using electrons, 4
    using light source to detect
        electrons, 5
    using waves of water, 3
Drude formula, 251
Drude model of metals, 171, 172

E

Effective mass, 40
Eigenfunction, 10, 14, 23
Eigenvalue, 10, 23
Einstein coefficient, 132
    stimulated and spontaneous
        emission, 183
Einstein, Albert, 7, 11
Electrode-induced potential, 90
Electromagnetic field energy density, 132
Electromagnetic perturbation of light, 126

Electron
    artificial design, 79–80
    band dispersions, 71
    characterization, 71
    correlation of solids, 51–55
    delocalization, 142
    diffraction, 60
    elementary charge, 100
    energy, 105–106
    exchange, 144
    index, 66
    indexing, 67
    one dimensional box, 15
    probability density, 32
    states of materials, 79–80
    transport, 75
    tunneling probability, 105–106
    wavevector, 75
Electron density of states, 69
    Fermi energy, 222
    Fermi level, 215
    materials with macroscopic
        dimensions, 65–66
    nanodots, 77–78, 79
    nanosheets, 67–71
    nanowires, 72–73, 74
    schematic illustration, 137
Electron density per unit length, 73
Electron orbitals
    hybridization, 214
    overlap, 144
Electron-exchange energy transfer,
        187–189
Electronic optical transitions, 133
Emission, 125–156
    spontaneous, 132
    stimulated, 131
Energy
    barrier, 92
    levels, 79
    potential, 230
    spectrum, 64–78
    transfer process, 185
Energy bands, 27
    formation, 38
    Kronig–Penny model, 35
Energy gap, 27
    formation, 32
    periodic potential, 32
Europium dibenzoylmethane complex, 198
Exciton. *See also* Frenkel exciton; Wannier
        (or Wannier–Mott) exciton
    Bohr radius, 135, 157

    charge-transfer, 134
    coherent length, 151
    displacement term, 147
    Hamiltonian operator, 147
    interactions, 133–134
    nanoscale materials optical properties,
        133–134
    phonon interaction, 155
    properties and description, 134
    radii, 153
    Rydberg energy, 135, 136
    superradiance, 154
Exotic localized electronic states, 97
External magnetic field, 213
Extraordinary Hall coefficient, 249
    extrinsic origins, 250
    side jump mechanisms, 250
Extraordinary Hall effect, 250

F

Fabry–Perot cavity, 194
Face-centered lattice, 41
Fermi distribution functions, 76, 106, 112
Fermi energy, 28, 30, 56
    DOS, 222
Fermi gas, 28
Fermi level, 112, 113, 214
    DOS, 215
Fermi liquid theory, 215
Fermi surface, 30
    reciprocal space, 30
Fermi wavelength, 65, 122
    valence electron, 58
Fermi's golden rule, 131, 132, 138, 153, 180
Fermi–Dirac distribution function, 215, 244
Fermi–Dirac statistics, 52
Fermion, 28, 51, 111
Ferrimagnetism, 208
Ferromagnetic domains, 59
    wall-related phenomena, 229–234
Ferromagnetic material, 224
Ferromagnetic nanoparticle, 228
Ferromagnetic spin system, 213
Ferromagnetic zigzag wires, 235
Ferromagnetic/nonmagnetic (FM/NM)
        interfaces, 247
    nonlocal detection of spin
        accumulation, 249
    spin accumulation, 247
Ferromagnetism, 208
    mean field theory, 211–212
    nanostructures, 222

Feynman's path integral approach,
    230–231
Feynman, Richard, 1
Finite potential wells
    graphical solution, 88
    numerical results of k and energy level
        of quantized states, 89
Fluorescence excitation spectrum, 184
Fluorescence lifetime, 187
FM. *See* Ferromagnetic/nonmagnetic
    (FM/NM) interfaces
Förster dipole–dipole energy transfer, 187
Förster distance, 184
Förster energy transfer, 59, 183
Förster radius, 185
Förster resonant energy transfer (FRET)
    nanoscale materials optical properties
        and interactions, 180–186
    symbols, 181
Free electron Fermi gas, 27–30
Free electron model, 31–32
Free-electron-like dispersion, 93
Frenkel exciton, 59, 133, 134
    nanoscale materials optical properties
        and interactions, 149–152
Frequency detuning, 130
FRET. *See* Förster resonant energy transfer
    (FRET)
Friedel oscillation, 122

G

Gallium arsenic quantum wells, 196
Giant magnetoresistance (GMR), 242
    effect, 59
    iron chromium, 239
    magnetic nanostructures, 240
    nanoscale materials magnetic and
        magnetotransport properties,
        240–244
    spin diffusion length, 245–248
GMR. *See* Giant magnetoresistance
    (GMR)
Gold
    calculated extinction, scattering, and
        absorption, 168
    extinction spectra, 175
    light wavelength, 170
    plasmon resonance, 175
    Q factor, 168
    SEM image, 175
Gold nanoparticles
    absorption spectra, 174

Coulomb blockade, 103
    ferromagnetic behavior, 223
Gold nanorods
    absorption spectra, 175
    TEM images, 175
Graphene
    band dispersion, 49
    energy band structure, 50
    tight binding approximation, 45–50
Graphite layer, 48
    atoms arrangement, 45
    band diagram, 45
    band dispersion, 50
    Brillouin zone, 49
Ground states
    nanoparticle, 215
    wave functions, 144

H

Hall effect. *See also* Extraordinary Hall
    effect; Spin Hall effect
    inverse spin, 251, 252
    nonmagnetic materials, 251
Hall resistance, 250
    ferromagnetic materials, 249
Hamiltonian matrix, 20
Hamiltonian operator, 13, 21, 23, 36, 128,
    131, 216
    dimer, 144
    excited state, 150
    exciton displacement term, 147
    hydrogen atom, 135
    interaction, 181
    single spins, 230
    spin system, 236
Harmonic function, 128
Harmonic oscillator, 15–16
Hartree's approximation, 27
Hartree–Fock approximation, 51–53
Hartree–Fock equation, 53
Heisenberg equation, 22
    spin system, 236
Heisenberg picture, 21, 22
Heisenberg quantum mechanics, 20–21
Heisenberg's equation, 18
Heisenberg, Werner, 9, 18
Hermitian linear operator, 21
Hermitian operator, 10
High symmetry, 43
Hund's rule, 207
Hydrogen
    electrons distribution, 220

energy level, 19
Hamiltonian operator, 135
magnetic field, 220
radial eigenfunction, 19
Schrödinger equation examples, 16–17

I

Icosahedron nanoclusters, 117
Indium arsenic quantum dots
    time-resolved photoluminescence
        intensity decays, 195
Indium gallium arsenic quantum wells
    time-resolved photoluminescence
        intensity decays, 196
Inelastic scattering, 113
Inelastic scattering processes, 114
Interaction Hamiltonian operator, 181
Interatomic bonds, 97
Interband transition energy, 139
Interference, 3
Inverse spin Hall effect, 252
    diagram, 251
Ionized aluminum magic numbers, 118
Iron
    calculated extinction, scattering, and
        absorption, 168
    DOS diagram, 242
    energy dispersion, 244
    Q factor, 168
    spin-up and spin-down states, 244
Iron chromium
    GMR, 239
    torque, 239
Iron/aluminum oxide/iron trilayer, 243
Isotropic energy, 116
Itinerant magnetism, 220

J

J-aggregate spectrum
    dipole–dipole interaction, 154
Jellium model, 92
Jordan, Pascual, 18
Jullire's formula, 244
Jullire's model, 242
    MR, 245

K

Ket vector, 18, 20
Kronig–Penny model
    energy band diagram, 35

solids electronic band structured,
    33–34
Kubo effect, 56, 59, 213, 218
    magnetotransport properties, 214–219
    nanoparticle, 219
Kubo formula, 251

L

Laguerre polynomial, 17
Lambert's law, 162
Landauer's formula, 242
    finite voltage, 244
Langevin–Debye theory, 212
Langmuir–Blodgett films, 198
Laplace's operator, 13
Lead
    Blügel's analysis, 222
    electron magnetic susceptibility, 219
    films magic thickness, 122
    films stability, 122
Legendre polynomial, 17
Light
    confined in cubic, 192
    confined in one, two, three
        dimensions, 193
    microcavity, 192
    wavevector values, 192
Light scattering theory
    geometry, 163
    nanoscale materials optical properties
        and interactions, 160–163
Light wavelength
    calculated scattering, 169
    calculated scattering and
        absorption, 170
    spherical gold particles, 170
Linear chain aggregate, 152
Lorentz absorption line shape, 126
Lorentz force, 249
Lorentz model of atom, 159
Lorentz oscillating dipole, 163
Lorentz oscillator, 160
Lorentz, H. A., 160

M

Macroscopic quantum coherence, 229
Macroscopic quantum tunneling, 229,
    231–232
Macrosize material size-dependent
    change, 96
Magic clusters, 117

Magic numbers
    extreme stability, 118
    ionized aluminum, 118
    sodium nanoclusters, 118
Magnesium oxide, 245
Magnetic anisotropy energy, 226
Magnetic domain wall, 232
Magnetic multilayers, 241
Magnetic nanoparticle, 228
Magnetic ordering, 208
Magnetization
    curves, 208
    landscape, 230
    macroscopic quantum tunneling, 232
Magnetoresistance (MR), 241
    Jullire's model, 245
Manganese atom corral, 82
Materials. *See also* Nanodots; Nanowires
    electronic properties, 57
    magnetic properties, 57
    nanostructures, 56
    optical properties, 57
    physical phenomena, 56
Matrix mechanics, 18–19
Matter wave, 7, 12
Matter–light interaction, 126
Maxwell relation, 161
Maxwell's equations, 128, 160
    electric field in vacuum, 192
    Mie solutions, 164
MBE. *See* Molecular beam epitaxy (MBE)
Mean field theory, 213
Mesoscopic, 58
Metal surface, 200
Metallic nanosphere, 101
Mie solutions, 172
    electric fields, 165
    Maxwell's equations, 164, 165
Mie theory, 164, 166
Mie's scattering, 159
Mie, G., 164
Modified damped Lorentz oscillator
        model, 171
Molecular beam epitaxy (MBE), 90
Momentum incident electrons at slits, 6
Monomer excited state, 148
Motion equation, 161
MR. *See* Magnetoresistance (MR)

N

N-spin system, 211
Nanoclusters magic numbers, 117

Nanodots, 77, 142
    electron DOS, 79
    energy levels, 79
    zinc sulfate antibunching
        phenomenon, 158
Nanomaterials, 96
    k-states, 73
Nanoparticle
    ensemble, 216
    Kubo effect, 219
    magnetic susceptibility, 218, 219
    quasiparticle excitation states, 215
    specific heat, 216–218
    thermodynamic properties, 216
Nanoscale materials electronic states and
        electrical properties, 63–124
    band dispersion effect, 93–95
    charging effect, 100–102
    edge (surface) localized states, 96–99
    electron DOS in 0D materials, 77–78
    electron DOS in 1D materials, 72–73
    electron DOS in 2D materials, 67–71
    electron DOS of 3D materials, 65–66
    electronic growth, 119–121
    electronically induced stable
        nanostructures, 115–121
    energy spectrum, 64–78
    finite potential well, 87–92
    lifetime broadening effect, 113–114
    low dimensionality, 64–78
    magic numbers in clusters, 116–118
    quantization, 79–95
    quantized conductance in 1D
        nanowire systems, 74–76
    quantized states shape effect, 84–86
    size effects limiting factors, 111–114
    space for electrons, 64
    thermal fluctuation, 111–112
    tunneling phenomena, 102–110
    two dimensional cylindrical well, 83–84
    two dimensional square well, 80–82
Nanoscale materials magnetic and
        magnetotransport properties,
        207–253
    current-induced domain wall motion,
        235–239
    current-perpendicular-to-plane (CPP),
        245–248
    electronic structures quantization,
        214–219
    exchange interaction, 208–210
    ferromagnetic domain-wall-related
        phenomena, 229–234

ferromagnetism mean field theory, 211–212
GMR and TMR effect, 240–244
GMR-spin diffusion length, 245–248
Kubo effect, 214–219
magnetic interface effect, 240–253
magnetic ions, 207
magnetic nanostructures macroscopic quantum tunneling, 229–232
magnetic ordering, 207
magnetism fundamentals, 207–212
quantum coherence, 233–234
single-domain structures, 224–228
spin accumulation, 245–248
spin Hall effect, 249–252
superparamagnetism, 224–228
surface effects, 213–228
three dimensional confined systems size effects, 213–228
transition noble metals surface magnetism, 220–223
Nanoscale materials optical properties and interactions, 125–204
absorption, 158–175
dielectric interfaces effects, 198–200
dielectric spheres size-dependent scattering, 164–168
electron-exchange (Dexter) energy transfer, 187–189
emission, 125–156
excitons, 133–134
Förster resonant energy transfer (FRET), 180–186
high-dielectric constant materials size effects, 136–139
light scattering theory, 160–163
linear optical transitions quantum mechanics, 126–132
local field enhancement, 176–179
microcavities optical interactions, 191–197
molecular excitons size effects, 153–156
molecular Frenkel exciton, 149–152
optical electrons finite number, 157
optical properties of metal nanoparticles, 169–175
particle–light interactions in finite geometries, 191–204
particle–particle size-dependent electromagnetic interactions, 179–189
photo-induced electron transfer, 190

pi-conjugated systems strongly interacting, 144–148
plasmonics, 169–175
radiative energy transfer, 179
scattering, 158–175
size-dependent optical properties adsorption, 125–156
surface-enhanced Raman scattering, 176–179
Wannier excitons, 135
Nanoscale square and rectangular quantum wells, 81
Nanosheet, 68
electron DOS, 70
k-states, 68
Nanosphere, 102
Nanosquare formation two dimensional, 81
Nanostructures
ferromagnetism, 222
formation, 64
size reduction, 64
Nanowires
channels contribution, 77
conductance, 75
conductance quantization, 77, 78
electron DOS, 74
electron transport, 75
electron wavevector, 75
Naphthyl
dimer, 149
transition dipoles, 149
Nearly free electron model, 27
Newton's second law of motion, 160
Newtonian mechanics, 12
NM. *See* Ferromagnetic/nonmagnetic (FM/NM) interfaces
Nonspherical Fermi surface, 44
Normalized system, 10

O

Oligophenylenevinylenes, 143
Oligothiophenes, 143
Orbital, 28
Orthogonal functions, 10
Orthogonal set, 10
Orthogonal system, 10
Oscillatory function, 84

P

Parallel dimer, 147
Paramagnetism, 208

Pauli exclusion principle, 28, 207, 215
Pauli paramagnetic susceptibility, 221, 222
Pauli susceptibility of metals, 218
Pauli, Wolfgang, 28
Perimeter-free electron orbital (PFEO)
    model, 140
Periodic electric field of light, 161
Perturbation theory, 22–26
Perylene
    derivative, 156
    dye molecules, 197
PFEO. *See* Perimeter-free electron orbital
    (PFEO) model
Phase accumulation rule, 95
Phenes absorption spectra, 142
PIC. *See* Pseudoisocyanine (PIC)
    toluenesulfonate dye
Planck, Max, 18
Plasmonics
    gold particles, 175
    nanoscale materials optical properties
        and interactions, 169–175
Probability amplitude, 5
Probability density, 8
Pseudoisocyanine (PIC) toluenesulfonate
    dye, 155
Pseudoisocyanine bromide J-aggregate, 156
Pseudoisocyanine chloride dye
    absorption spectra, 154
    J-aggregate form, 154
Purcell effect, 195
Purcell, E. M., 195
Pyrene donor molecular, 187

Q

Quantized conductance in nanowire
    systems, 74–76
Quantized states
    energy, 86
    leaky boundaries, 114
    lifetime broadening, 114
    numerical results, 86
Quantum chemical molecular orbital
    methods, 140
Quantum conductance, 77
Quantum confinement, 83
Quantum dots, 77, 142
    zinc sulfate antibunching
        phenomenon, 158
Quantum fluctuation, 230
Quantum mechanics and band
    structure, 1–60

    laws, 66
    nanostructures properties, 56–69
    solids electronic band structured, 27–55
Quantum mechanics fundamentals, 1–26
    eigenvalue, 10
    electron in one-dimensional box, 14
    expansion theorem, 10
    expected value, 10
    harmonic oscillator, 15–16
    hydrogen atom, 16–17
    interference effects, 1–4
    matrix mechanics and bra-ket (Dirac)
        notation, 18–19
    operators, 8–9
    perturbation theory, 22–26
    probability amplitude effects, 1–4
    Schrödinger equation, 11–12, 14–17
    superposition principles, 13
    uncertainty principle, 5–6
    wave functions, 7
Quantum tunneling, 231
Quantum well states, 94, 112
    manganese atom corral, 82
    parabolic and nonparabolic dispersion,
        95, 96
    shape effect, 85
Quantum wells
    absorption spectra, 138
    confinement potential, 87
    finite potential barrier, 89
    semiconductor structure, 138
Quasiparticle excitation states, 215

R

Radiative energy transfer, 179
Raman scattering, 177, 178. *See also*
        Surface-enhanced Raman
        scattering (SERS)
Random matrix theory, 216
Rayleigh scattering, 162–164, 164, 166
Reciprocal lattice, 30
    vectors, 49
Reciprocal space, 30
Resonant energy transfer
    between donor and acceptor, 180
    distance dependence, 188
Resonant interaction, 148
Rhodamine, 178
Riccati–Bessel functions, 166
Ruthenium monolayer spin
    polarization, 224
Rydberg exciton energy, 136

S

Scaling laws, 170
Scanning tunneling microscope (STM)
        image, 69, 108–112
    apparatus, 109
    atomic resolution, 82
    energy diagram for tunneling
        current, 110
    silver films, 120
Scattered light intensity
    calculated spatial distribution, 167
    spherical gold particles water
        suspension, 167
Schrödinger equation, 13, 83, 127, 129,
        141, 214
    dimer ground state, 144
    electron in one-dimensional box, 14
    examples, 14–17
    harmonic oscillator, 15–16
    hydrogen atom, 16–17
    quantum mechanics fundamentals,
        11–12
Schrödinger notation, 19
Schrödinger picture, 20, 21, 22
Schrödinger quantum mechanics, 20–21
Schrödinger's cat
    thought experiment, 6
Schrödinger, Erin, 6, 11
Selenium quantum dot
    antibunching phenomenon, 158
    fluorescence time trace, 158, 159
    time delays histogram, 158
Semiconductor
    materials, 137
    nanocrystals, 140
    nanostructures, 142
    optical transition to excitonic states,
        136
    structure, 138
SERS. *See* Surface-enhanced Raman
        scattering (SERS)
Shockley states, 100
Silicon
    dual-bias STM image, 111
    energy band diagrams, 44
    silver surface, 111
    solids electronic band structured, 44
Silver
    Bohr magneton, 223
    calculated extinction, scattering, and
        absorption, 168
    dual-bias STM image, 111

    magnetic moments, 223
    nanofilms, 70
    Q factor, 168
    sphere, 177
Silver films
    morphology, 120
    STM images, 120
    two-step method, 120
Simple lattice, 41
Single-domain nanoparticle, 227
Size effects
    calculated scattering, and
        absorption, 169
    concepts, 63
    molecular excitons, 153–156
    nanoscale materials electronic states
        and electrical properties, 111–114
    nanoscale materials magnetic and
        magnetotransport properties,
        213–228
    nanoscale materials optical properties
        and interactions, 136–139
    pi-conjugated systems, 140–143
Slater determinant, 209
Sodium nanoclusters magic numbers, 118
Solids electronic band structured, 27–55
    Block function, 33
    Brillouin zone, 40–43
    density functional method, 54–55
    effective mass, 37–39
    electron correlation, 51–55
    free electron Fermi gas, 27–30
    graphene tight binding approximation,
        45–50
    group velocity, 37–39
    Hartree–Fock approximation, 51–53
    Kronig–Penny model, 33–34
    nearly free electron model, 31–32
    phase velocity, 37–39
    quantum mechanics and band
        structure, 27–55
    reciprocal lattice, 40–43
    silicon energy band structure, 44
    tight binding model, 35–36
Spatial distribution, 92
Spherical metallic nanoparticle, 173
Spillage-induced interface dipole
        layer, 121
Spin accumulation, 247–248
Spin current, 236
    induced magnetization switching, 240
Spin filter, 245
Spin glass, 208

Spin Hall effect, 59, 240
    diagram, 251
    experimental confirmation, 252
    nanoscale materials magnetic and
        magnetotransport properties,
        249–252
Spin transfer torque, 236
Spin wave functions, 189
Spin-orbit coupling, 217, 218
    Hamiltonian operator, 250
Spontaneous emission, 132
Spontaneous magnetization, 222
Square well potential, 214
Stationary state, 20
Stern–Gerlach experiments, 223
Stimulated emission, 131
STM. *See* Scanning tunneling microscope
        (STM) image
Stoner criteria for ferromagnetism of
        itinerant electron systems, 222
Stoner enhancement factor, 222
Subbands, 69
Superradiance, 156
Surface-to-volume ratio, 96
Surface-enhanced Raman scattering
        (SERS), 160, 177
    nanoscale materials optical properties
        and interactions, 176–179
Surface-localized electronic
        states, 97
    spatial distribution, 100
Symmetric phenes, 142

T

Thermal broadening width, 112
Thought experiment, 1
    Schrödinger's cat, 6
Tight binding model, 27, 93
    energy bands formation, 38
    solids electronic band structured,
        35–36
Time-averaged Poynting vector, 163
Time-dependent Schrödinger equation,
        126, 128
TMR. *See* Tunnel magnetoresistance
        (TMR)
Total exchange energy Hamiltonian
        operator, 211
Transition dipole moments, 148
    electronic transitions, 129

Tunnel magnetoresistance (TMR), 59, 242
    nanoscale materials magnetic and
        magnetotransport properties,
        240–244
    ratio development, 245
Tunneling
    gap of STM, 103
    macroscopic quantum, 229, 231–232
    phenomena, 102–110
    probability of electrons, 105–106
    process, 243
    thermal activation, 231
Tunneling current, 104
    distance dependence, 107
    energy diagram for calculation, 106
Two-band theory, 99

U

Ultrathin lead films magic thickness, 122
Ultrathin metal films magic thickness, 121
Uncertainty principle, 5–6
Uniaxial anisotropy, 226
Unpaired bonds, 98
Unperturbed eigenvalues, 23
Unperturbed Hamiltonian operator, 127

V

Valence electrons
    confined, 91
    Fermi wavelength, 58
Van der Waals interaction, 145

W

Wannier (or Wannier–Mott) exciton, 59,
        133, 134, 139, 151, 153
    confinement, 126
    nanoscale materials optical properties
        and interactions, 135
Wave functions, 32, 80
    ground states, 144
    quantum mechanics fundamentals, 7
    spin, 189
Wavevector
    electronic structure, 29
    light values, 192
Weiss field, 213
Wentzel-Kramers-Brillouin (WKB)
        approximation, 108

Wigner–Seitz cell, 41, 42
    two dimensional, 42
WKB. *See* Wentzel–Kramers-Brillouin
        (WKB) approximation
Work function, 92
    surfaces, 93

Z

Zeeman energy, 212
Zeeman splitting, 220
Zinc sulfate quantum
        dot, 158

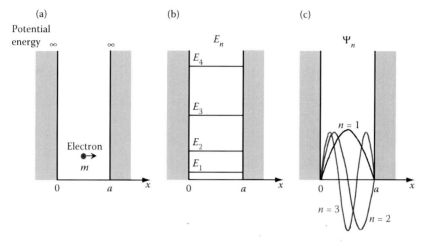

Figure 1.5 An electron in a 1D box: (a) potential energy, (b) energy levels, and (c) wave functions.

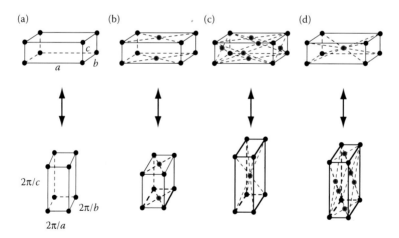

Figure 1.14 (a) Simple lattice, (b) cubic-centered lattice, (c) face-centered lattice, and (d) body-centered lattice.

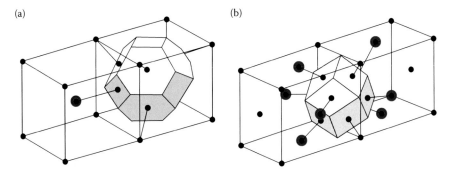

Figure 1.16 (a) The Wigner–Seitz cell of a body-centered cubic (bcc) structure and (b) the Wigner–Seitz cell of a face-centered cubic (fcc) structure.

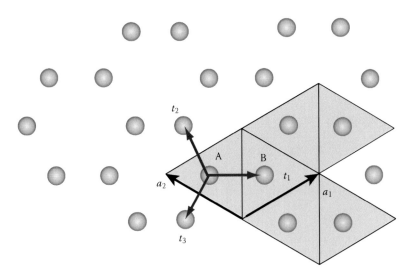

Figure 1.19 Arrangement of atoms in a graphite layer.

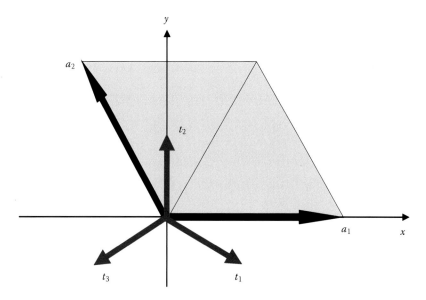

Figure 1.20 Unit cell of a graphite layer.

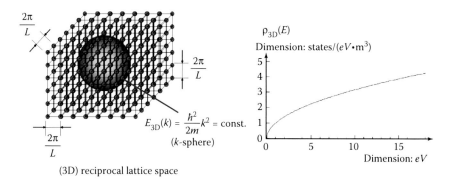

$E_{3D}(k) = \dfrac{\hbar^2}{2m}k^2 = \text{const.}$

(k-sphere)

(3D) reciprocal lattice space

$\rho_{3D}(E)$

Dimension: states/($eV{\cdot}m^3$)

Dimension: eV

Figure 2.3 The Fermi sphere in the 3D reciprocal lattice (left) and the electron DOS (per energy) of the 3D crystalline material (right).

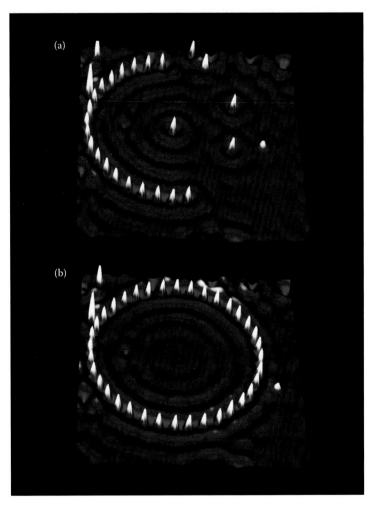

Figure 2.17 Construction of a 2D cylindrical well using STM (a) during the construction and (b) the completed cylindrical well. (Reprinted from S. Hla, K. Braum, and K.H. Rieder, *Phys. Rev. B* 67, 2003, 201402. With permission.)

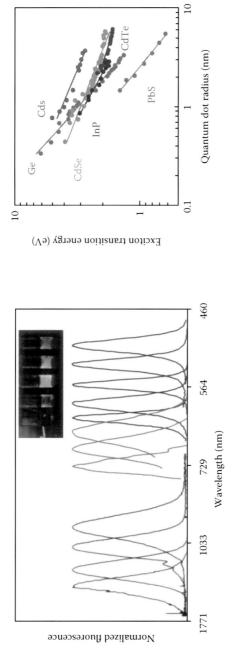

Figure 3.8 Left: Size- and material-dependent fluorescence spectra of semiconductor nanocrystals. Blue lines: CdSe nanocrystals with diameters of 2.1, 2.4, 3.1, 3.6, and 4.6 nm (from right to left). Green lines: InP nanocrystals with diameters of 3.0, 3.5, and 4.6 nm. Red lines: InAs nanocrystals with diameters of 2.8, 3.6, 4.6, and 6.0 nm. Inset: Images of a series of silica-coated core (CdSe) shell (ZnS or CdS) nanocrystals. (Reprinted from M. Bruchez et al., *Science* 281, 1998, 2013. With permission.) Right: Size-dependent exciton energies for quantum dots of various semiconducting materials. (Reprinted from G.D. Scholes and G. Rumbles, *Nat. Mater.* 5, 2006, 683. With permission.)

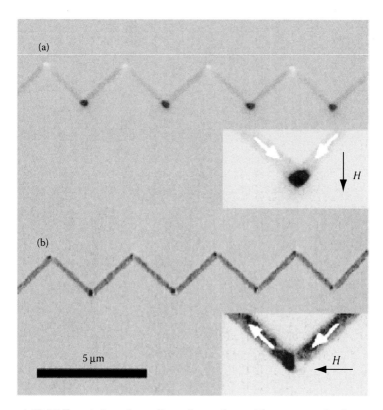

Figure 4.18 Different domain wall configurations of ferromagnetic zigzag wires after applying magnetic field (a) perpendicular and (b) parallel to the horizontal axis.

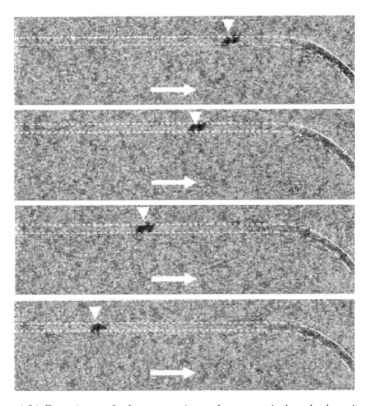

Figure 4.21 Experimental demonstration of current-induced domain wall motion. (Reprinted from A. Yamaguchi et al., *Phys. Rev. Lett.*, 92, 2004, 077205. With permission.)

(a)   (b)

n_s (a.u.)   Reflectivity (a.u.)

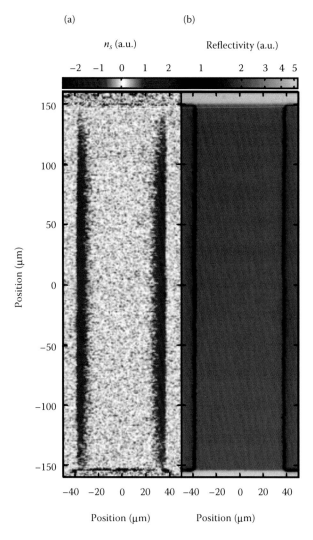

Figure 4.34 Experimental confirmation of the spin Hall effect. (Reprinted from Y.K. Kato et al., *Science*, 306, 2004, 1910. With permission.)